Quality in Practice

JOHN A. MURPHY

WITH

TONY FARMAR

GILL AND MACMILLAN

Published in Ireland by
Gill and Macmillan Ltd
Goldenbridge
Dublin 8
with associated companies in
Auckland, Dallas, Delhi, Hong Kong,
Johannesburg, Lagos, London, Manzini,
Melbourne, Nairobi, New York, Singapore,
Tokyo, Washington
© John A. Murphy, 1986
Reprinted 1988
7171 1462 7
Print origination in Ireland by Printset & Design Ltd., Dublin

Printed by Criterion Press, Dublin

All rights reserved. No part of this publication may
be copied, reproduced or transmitted in any form or by
any means, without permission of the publishers.

*To my wife Eleanor
and children
Niall Augustine, Maria and John Gerard*

Contents

	Acknowledgements	ix
1	Introduction to quality management	1
	Part I: Right First Time	5
2	Quality starts at the top	7
3	Product design	18
4	The buying function	27
5	In-process control	36
6	Good Manufacturing Practice	47
7	Just-In-Time production systems	57
8	Information quality standards	64
9	The customers' contribution to quality	71
10	Service without a snarl	77
	Part II: Pride in the product	85
11	Making the system work for quality	87
12	Motivation — Getting to Yes	95
13	Putting your point effectively	105
14	Training for quality	114
15	Quality circles	122
	Part III: The quality toolkit	131
16	The quality manual	133
17	Quality auditing	142
18	Metrology	150
19	Standards	157
20	Problem solving	166
21	Letting statistics help quality control	178
22	Inspection and sampling	191
23	Documentation and control charts	200
24	Computers	209

	Part IV: The Quality Environment	219
25	The economics of quality	221
26	Getting close to the customer	228
27	Consumer protection	236
28	Product liability	242
	Appendix 1: EEC Directive on strict product liability	248
	Appendix 2: Sources of further assistance	255
	Appendix 3: A typical quality audit checklist	259
	Bibliography	262
	Index	265

Acknowledgements

FEW BOOKS today contain solely the input of one person. In writing this book I have benefited from the experience and knowledge of many people and organisations. I am especially grateful to the management of R. & A. Bailey & Co. Ltd, notably D. I. Dand, K. MacCarthy Morrogh and T. M. Murray for their considerable support. Their unselfish assistance and co-operation were invaluable and are acknowledged with sincere gratitude. I am also indebted to the following organisations and people for their assistance:

AnCO — The Industrial Training Authority — G. Kelly; Ballyclough Co-Op Creamery Ltd — J. Gleeson; Ballyfree Farms — Dr P. Griffin; Braun Ireland Limited — B. Hodge; P. Carney Ltd — J. Flood; Chemoflon GmbH — D. Keogh; Dairyland Ltd — J. O'Sullivan; Dataproducts (Dublin) Ltd — Dr R. Wilson; Digital Equipment International Ireland B.V. — T. Fernandes; Drogheda & Dundalk Dairies — J. Callery; Eurolift Ltd — B. Cole; FMC International — D. O'Grady; Howmedica International Inc. — B. Penn; Institute for Industrial Research & Standards — B. Rothery; Irish Fher Laboratories — Dr C. O'Brien; Kilkenny Design Workshops — J. Dunne; Measurex (Ireland) Limited — M. Hanratty; Merck Sharp & Dohme (Ireland) Ltd — J. R. Kelly; Mitchelstown Co-Op Creameries — S. O'Connell; NCF — Sligo Dairies Ltd — F. McKiernan; NEC Ireland Ltd — K. Moreland; Nypro Ltd — T. Fitzgerald; Oriflame Manufacturing Ltd — N. Hitchcock; Pfizer Chemical Corporation — Dr J. C. Riordan; Semperit (Irl) Ltd — V. Marsland; System Industries Europe — J. A. Neary; Thermo King Europe — S. Fordham; Virginia Milk Products Ltd — D. Courtney; Wang Laboratories Ireland B.V. — D. Collison; Wavin Ireland Limited — J. A. Brownlee; Westinghouse Electric Systems & Logistics Limited — M. Malone.

Among those who gave generously of their time in reading the initial typescript and offering constructive criticism were: Dan Boland, Frank Curley, Pat Given, Jim Murray, Tommy Murray and Michael Wallace. Particular appreciation must be given to Tony Farmar who has been deeply involved in all aspects of planning, developing and editing of the book. Thanks are also due to Gill and Macmillan Ltd for their help and assistance. Finally I extend my deepest gratitude to my wife Eleanor for her ongoing support. Without her constant help, patience and encouragement over many years, this book would not exist.

Chapter 1

Introduction to quality management

IRELAND is famous throughout the world for its beer and for its butter — but it was not always so. Both these products achieved their present pre-eminence because of a positive decision by individuals to put the quality of the product above all else.

In the 1820s the Irish economy was in decline, and the sales of Guinness, only one of several large breweries at the time, were falling. By 1824 they had dropped to what they had been at the beginning of the century. The firm entered its darkest period, but the then head, Arthur Guinness II, made a momentous decision. Instead of watering down the beer as his fellow brewers were doing, he decided to produce a better stout than anyone else. Exact regulations were laid down for 'Extra Superior Porter', with the best materials, more hops and stricter quality control. The policy worked. His fellow brewers and their watered beer fell by the wayside, and by 1833 Guinness had achieved a dominance in Ireland it has never lost since.

The story of Irish butter is less cheerful. As the British market for dairy products grew in the nineteenth century, travelling butter-buyers would come round the agricultural districts. They would set up in the stables or the pub, and the men of the parish would bring the week's butter. As a contemporary put it, the butter-buyer 'would welcome them, would praise the butter, would praise the women, he will give the highest price to each of them. He will find no fault with greasy butter, nor with ribbed butter, nor with butter not properly washed from buttermilk, nor from smoky butter ...' As a result of this arrangement, a firkin of Irish butter arriving on the British market would frequently contain layers of butters of varying ages, flavours, colours and textures, not to mention aromas. It was not surprising that the Danes, the Swedes and the French beat the Irish out of the British butter market until Kerrygold was launched in the late 1960s.

Nowadays, the dairy industry has been totally converted to the quality approach. Sligo Dairies, for instance, report that 'as a result of the awareness of quality and a quality system, the Dairies' sales of liquid milk in the Mayo region went from 5 per cent share in 1979 to 70 per cent in 1985'. At the same time the number of dairies competing in the region has doubled from four to eight. The levels of returned product due to handling, bacteriological faults or over-ordering are now less than 0.5 per cent of sales as opposed to an industry average of 3–5 per cent.

Quality in Practice

R. & A. Bailey's plant, Western Estate, Dublin   R. & A. Bailey

The anecdotal evidence of these stories is supported by the findings of the PIMS research. The findings of this Harvard University based database on successful and unsuccessful companies are discussed in detail in Chapter 25, 'The economics of quality', but it is worth outlining them here as well. The database consists of over two hundred separate pieces of information on over 3,000 companies. The information ranges from financial data such as return on capital, profit levels, stock turnover, etc., to details such as degree of unionisation. The purpose of the data is to discover which factors make companies successful and what makes them fail. The answer is very clear. The best way to achieve success is to have a monopoly in a rising market. This is unfortunately not to be had just for the asking. The second most important factor in achieving consistently high return on capital was what the database calls *Relative Product Quality*.

Relative product quality is not defined according to technical or engineering standards, nor is it a function of this year's quality compared to last. Relative product quality is solely judged
— from the customer's point of view, not the company's
— by both the product and the associated services
— relative only to competitors in the served market
— without reference to price.
In short, relative product quality is based on everything about the product except the price. It is this definition of quality that is used in this book. We are not therefore saying that a Saab is necessarily of better RPQ than a Fiat Ritmo. For the market it is aimed at, either may provide much greater quality than the other.

The conclusions from the PIMS database declare that a high relative product quality produces more long-term profits than a high expenditure on marketing, more than heavy investment in capital equipment, and is nearly as valuable as a high market share. Most Irish firms are small, and therefore cannot expect to command a high market share. They should therefore, according to the PIMS theory, concentrate the bulk of their strategic effort on improving their relative product quality. But a recent in-depth study by University College, Galway of six small firms showed that only one of them (which was in the food trade) had any formal quality control system at all. The case study records a typical example.

Exhibit: A small engineering firm
Products: various engineering products in stainless steel, mild steel and aluminium; cabinets, metal cases for the electronics industry; meat hooks and systems for the meat trade; cleats, angle-plates, tie-rods, etc. for the construction industry; 'U' bolts, spare wheel carriers, etc. for the motor industry.
Numbers employed: 20
Manufacturing procedures: A specification drawing is usually available for each new product. The drawing is examined, and if there is any doubt about the specification, the customer is contacted. The production foreman verbally instructs each worker on the task to be performed by him. Each worker is responsible for the quality of his work. Errors, when made, are identified by the production foreman. There are no fixed inspection stages. When appropriate, samples are made before production commences. Customer complaints are recorded and filed.
Source: University College, Galway

The study suggested that these six firms were typical of the 95 per cent or so of Irish firms that employ less than 100 people. Generally they lacked a single person responsible for quality control (notice the optimistic remark 'Errors, when made, are detected by the production foreman'); work instructions were always verbal; any quality control relied on inspection rather than 'right first time' attitudes; production rather than quality was the major concern; quality control as such was regarded as obstructive and expensive.

No one quality assurance system could ever be suitable for every company. The printing of magazines clearly requires quite different controls from the production of gourmet meals in a restaurant. But the underlying principles are the same, and go back to the cycle of control discussed in Chapter 11. Once the concept of control is employed, only the specifics of the situation change.

Quality consciousness is not something that can be bought in on Monday. It is an important development in what is called the 'corporate culture'. This is the collection of ideas that everyone in the company has about what they are doing and why. Some companies are obsessed with profits, almost to the exclusion of the product, while others are run by engineers who consider the production function first, last and always, to the exclusion of the customer.

The managers' views of what is important form a significant part of corporate culture. If the chief executive is an accountant, he will naturally be primarily concerned with return on capital, with budgeting and with costs. Product quality will be left to the factory people, who are supposed to know about such things. It is unfortunately just this kind of well-meaning manager that gets business a bad name with the public. After all costs are not the customers' problem. What they want are well made (and preferably Irish-made) products above all. The companies that have a good image with the public are those that have proved they are concerned about the quality of the delivered product. An Irish survey in 1983 discovered that 45 per cent of consumers, when asked 'What do you think is the most important feature when you buy a product?', listed quality. And when they said quality, they did not mean 'the best', as a later question proved, when 67 per cent picked out 'good value' or 'genuine' as what they meant by quality.

The quality philosophy in action puts the making of the product the customer wants first. Research has shown that this not only increases the return on investment in the long run, but also improves the public image of the company. This in itself has an important motivating effect on the staff. It is much more satisfying to work for a company that other people respect than for one they do not. This book is an introduction to quality in practice. It is divided into four parts. The first describes how the quality approach affects various departments throughout the firm, from top management, through design, right out to the various service departments; the second part describes the all-important aspects of motivation, communication and training; the third part deals with the quality disciplines, those skills and documents that may not yet be present in the company before launching a quality drive; the fourth part discusses the quality environment, the economics of quality, market research, the consumer movement and, finally, the very important forthcoming product liability law.

PART I
Right First Time

Chapter 2

Quality starts at the top

EVERY company in the world claims to provide a quality product. 'We spend a fortune', they say, 'on quality inspectors and quality control systems. The Quality Control Manager reports direct to the Production Director.' In other words, quality is seen firstly as a production problem, and secondly as a problem that can be handled by inspection processes. Unfortunately that is no longer enough. To survive in the new legal and market environments senior management have to attend to questions of quality management in detail. There are three pressing reasons.

1. *Product liability*: As the markets in the developed world become more sophisticated, customers expect more. Consumer movements are becoming more demanding, and national legislatures are responding to these demands. Very soon manufacturers in all EEC countries will be strictly liable in law for any damage or injury caused by design or manufacturing defects in their products, just as they are now in America. This new law will remove the need for injured parties to prove any negligence on the part of the manufacturer. Inevitably this will involve the production in court of batch inspection records, inspection sampling schemes, quality manuals and all the apparatus of a sophisticated quality assurance scheme. This 'strict product liability' is discussed further in Chapter 28.

2. *Relative product quality*: More positively, recent research into business success and failure has highlighted the importance of the relative quality of products. The better the quality of the product, the higher the return on investment the company will achieve. Relative quality is more important than relative price. Furthermore, each product should be compared only with those it competes with directly. We shouldn't scorn a daisy because it isn't a rose. And most importantly, that perception, both as to competitors and as to significant qualities, is as perceived by the customers, not by specialists on design award panels. Winning companies are those that discover, by market research and by listening to the customers, what is wanted, and then provide those elements better than anyone else. On the other hand they don't waste time and money providing high-quality aspects that the customers don't want.

3. *Quality costs*: The third element that necessitates the involvement of senior

management in quality assurance is the implications of the quality cost concept. This is a new grouping of cost elements that highlights the surprisingly large sums ordinary companies spend on quality and the failure to achieve it. Some of these cost elements, such as the cost of scrap or the wages of quality control inspectors, are readily available. Others, such as the quality planning elements in design and testing or the cost of complaint handling and adjustment, are less easy to unravel from their departmental budgets. When the four elements
— internal failure costs (rework and scrap, etc.)
— external failure costs (customer complaints handling, cost of returns and replacement, warranty costs, etc.)
— appraisal costs (inspection and testing, etc.) and
— prevention costs (quality planning and training, etc.)
were added together, they could often add to as much as 20 per cent of total sales. If even a quarter of this cost could be cut down, 5 per cent of total sales would be added straight to profits. Quality costs are described in detail in Chapter 25.

When companies began to investigate how quality costs could be cut, they came across a paradox. The more time and money they invested in quality planning, the lower their overall quality costs. The more they spent, the more they saved. Total quality costs could be reduced by as much as 35 per cent by investment in a fully working quality assurance system. Companies found that prevention by planning was cheaper than cure by inspection. Not only that but the relative product quality of the output went up, giving the company the strategic benefits of a good position in the marketplace.

At the moment Japanese companies dominate the colour TV market. They do so because, although the picture quality and general design of their sets are very similar to their competitors', they are much more reliable. One estimate suggests that in the late 1970s Western products had between two and four times the failure rates of Japanese ones. This is mainly because the Japanese spend more time solving the quality problems before they go into the marketplace. A typical schedule for the sources of field failures would be:

Source of problem	Proportion of failures
Development and design	20-40%
Quality of components	40-65%
Quality of workmanship	15-20%

These problems are all solvable, mostly in advance. Designers can be trained to give major weight to quality and reliability considerations as well as technical and design factors. Supplier assessment programmes can be designed to give a much stricter control of incoming materials. Quality of workmanship can be upgraded by training of operatives, by well designed process and process control systems, and by good inspection and test routines. Virtually all these cures require an initial investment in planning and training by the company to achieve the desired effect.

Initiating a quality programme

However, for senior management merely to recognise the desirability of quality assurance is not enough. Certainly something positive has to be done, but the proposed changes are so wide in implication, it is often difficult to know how to start.

This problem is accentuated by the typical background of senior managers. Very often they have marketing or finance training, but little or no knowledge of engineering and the technical aspects of production processes. As a result they may fail to gain a deep understanding of information presented in technical form. This was the origin of the quality cost idea: if quality problems and solutions could be presented in the financial language of senior management, the importance of quality considerations might be brought home.

There are of course other difficulties. Culturally, production controllers and shop-floor workers do not like senior management 'interfering' in the existing arrangements. Very often there is an almost equal resistance on the part of management to risk making fools of themselves. In many offices, for instance, a major limitation on the use of computers is the reluctance of 'old dogs' to learn new tricks in public. Male managers are reluctant to do any serious keyboarding, drastically limiting their access to the new computing power. Any change in the operating environment has to be introduced carefully, with the active participation of the shop floor and production management. The changes should start on a small scale, and gain acceptance gradually.

To break this cycle of resistance, senior management have to initiate a quality drive. This will probably run as follows:

Identify the existing quality levels

The first stage in going anywhere is to discover where you are to start from.

1. *Decide what the key quality characteristics are*: These will determine the quality of the product in the eyes of the customer. This may involve the use of market research or a special information gathering assignment for salespeople. A detailed analysis of customer complaints can be very revealing, particularly when contrasted with internal quality ideas. For instance if the company believes that colour is critical, and all the customers' complaints are about texture, then something is obviously wrong! An important objective of this exercise is to establish a clear view of the circumstances in which the product is used: for instance the manufacturer of fire safety equipment has to provide instructions that can be read and understood in seconds, not an elaborate sixteen page booklet on all the subtle advantages of the equipment.

2. *Assess the product's relative product quality*: This is done by grading the quality characteristics of the product in order of importance. The best analysis will come

Marketing has a key role to play in achieving high quality   R. & A. Bailey

from a group with representatives from both production and marketing. Include every possible characteristic about the product that may be of concern to the customers, except price. Each attribute should be assigned a weight or rating, so that the total of the weights adds up to one hundred. Thus a tissue manufacturer might assign high weights to box decoration and wet-strength, and low weights to precision of trimming or box cardboard quality. Ideally the weights should be derived directly from the customers by market research. The group should then assign scores (0 to 10) for each characteristic both for their own product and, because this is an investigation into *relative* product quality, those of the competition. In some cases the firm's product will be clearly superior, in others inferior to the best on the market. If the group can't assign a score, assume that the characteristic is average for the market.

Exhibit: Calculating a relative product quality table

Characteristic	Weight	Brand A Score	Brand A Value	Brand B Score	Brand B Value
Design	20	3	60	5	100
Ease of use	15	5	75	5	75
Speed of delivery	10	7	70	3	30
Reliability	35	8	280	6	210
Maintainability	5	5	25	5	25
Durability	15	4	60	7	100
Totals	100		570		540

In this comparison, Brand B scores high on design and durability, but the weights given to these characteristics are much less than those given to reliability, which is Brand A's forte. Brand A therefore has the higher relative product quality. On the other hand, if the group had decided that design was valued more highly by the customers than reliability, the situation would be reversed.

This exercise will highlight those areas in which the product's quality is relatively high, and those in which it is inferior to its competitors. By relating the score for a characteristic to the customers' sense of priorities, the company can start its quality assurance programme confident that the critical quality problems are being tackled first.

3. *Discover who establishes the standards of quality*: Willy-nilly the company is working to some standard. It may be merely a level of workmanship learnt during apprenticeship, tempered by complaints from salespeople, or it may be more detailed. For each of the key quality characteristics discover who sets the standards.

4. *Discover how the standards are applied and enforced*: This is very much *not* a matter of discovering how the quality inspectors work. The only way to create a consistent quality product is to make it right first time. So this stage investigates the points in the production process where the key quality characteristics are produced. These are the 'quality control points'. The individual workers and their supervisors at these critical points are the ones who will make or break the success of the quality plan. If the right standards are enforced at these points, then the outgoing quality levels will improve.

Establish new quality objectives

The next stage in the journey is to decide exactly where we want to go. As usual in life, this is a matter of desire tempered by cost.

1. *Costs*: The discovery of the full costs of quality (or lack of it) can be alarming. When the obvious elements such as inspection and rework are added to the less obvious ones such as problems caused by bad design or by the low quality image of the product in the marketplace, the reduction of these costs can become a major corporate objective. The way to reduce them is to design and make the product correctly in the first place.

2. *Standards*: Company quality standards should be set, detailing the long-term quality objectives of the company and how these are to be achieved. Specific targets and standards should be written down for each part of the process, and the various service departments such as customer complaint handling. These will be based on the list of key quality elements, so that everyone in the company knows exactly what is expected of them. Methods of achieving and maintaining these standards (through training, etc.) should also be specified.

Exhibit: Statement of company quality policy, R. & A. Bailey
In building and sustaining a brand in fiercely competitive international markets we have come to understand that Quality Assurance is a most important instrument of marketing.

The quality of our product is determined first and foremost by the consumers. The responsibility for discovering what the customer wants rests with marketing; assuring the brand is produced in a manner that provides this quality is the responsibility of the organisation at large.

To achieve our quality objectives, we must develop and foster an environment for the assurance of quality. Our quality philosophy should never tolerate an 'it's not my job' attitude. It is all of our jobs. We must encourage this
— by clear and decisive management support
— by accepting that quality assurance is a company-wide responsibility and therefore each employee in the organisation has a role to play in the achievement and maintenance of quality standards
— by providing the necessary training and other means to transmit and encourage this commitment to quality.

The techniques of quality control are the means by which we provide the assurance that quality is delivered. The specifications and operating procedures in the Quality Manual and the various specialist techniques are essential disciplines. But Baileys Irish Cream will only retain its pre-eminence if everyone working in the company sincerely wants to deliver quality to the customers.

When this attitude pervades a company, at least half the marketing job has been done. Without it, the other half is not worth doing.
Source: R. & A. Bailey

---

3. *Objectives*: Supplementing this long-term plan should be 'annual quality objectives'. These are specific targets decided by the group involved (managers, supervisors, etc.). Typical examples would be phrased like this:
— reduction in reject proportion by ...%
— reduction in inspection costs by ...%
— improvement in suppliers' average delivery speed by ... days per order.

Design
— concept etc (p18)

Ensure that products perform as stated
(+ safe + reliable)
Design Documentation — Product Safety
COMPLETE / RETRIEVABLE / ACCURATE

- USABILITY (p20)
- SAFETY
- RELIABILITY
- MAINTAINABILITY
- TRACEABILITY

⟹ chapter 8

INFORMATION Q' STANDARDS

— are messages understood
(Context a multi-natural consortium)
(Source, encoder, message, channel, decoder, recip)
p64

26/4/9;-
Quality Control Problems (p128) Simple.
— how things go wrong

The Quality Manual p133 (→)
Audit (feel 'Info Gathering) p146

17/4/85 Quality   IS ↓ IS NOT
                        INCLUDE
                              REMOVE

Altchamire:

(i) Existing Quality Levels (R9)
  (a) What are the key quality
                    characteristics?
      (ie Customer feedback/Complaints)
  (b) What is the RPQ (Relative Pr.Q)

Quality Characteristic — who sets
             its standard?
Make it right ∴ investigate the
points in the process where the key
characteristics are produced.

○ — what mechanisms must
we identify in order to make x̄

                    results
MOTIVATION   ⇒ ○ ⇒   CUSTOMER
INVESTMENT   RECALL FEEDBACK
ORGANISATION   Eg

(ii) Q Project Team — detailed Q Eng
     — Set targets + reports

As each individual objective is achieved, it becomes part of an overall quality improvement. The single target concentrates the mind, and avoids the temptation to make rousing but vague speeches about how quality is everyone's business. Everyone's business is no one's responsibility.

4. *Quality assurance scheme*: Simultaneously senior managers should be developing the quality assurance scheme. Among other things, this involves developing a quality manual (see Chapter 16), an investment in quality training (see Chapter 14), setting standards (see Chapter 19) and organising a quality audit schedule (see Chapter 17).

Exhibit: The elements of a quality system
The Irish Standard on quality management (IS 300) identifies the requirements for a fully fledged quality system. These are:
A. Control of pre-production quality
  1. Management responsibilities: policy, organisation, etc.
  2. The quality system
  3. Contract review
  4. Design, development and specification control
  5. Documentation control
  6. Purchasing
  7. Item identification and traceability
B. Control of production quality
  8. Production process control
  9. Inspection and testing
  10. Control of manufacturing and test equipment
C. Control of post-production quality
  11. Handling, storage, delivery and installation
  12. Inspection and test status
  13. Control of non-conforming items
  14. Corrective action
  15. Quality records
  16. Quality audits
  17. Training
  18. Marketing and servicing
  19. Number and frequency of assessments/inspections and tests performed
Source: IS 300

Implementing the quality assurance scheme

Although the description of every company's activities can be reduced to the same general terms (input, output, costs, results, etc.) in practice each is different, and so the actual implementing of a quality plan can only be described in generalities. Three elements are important. These are: motivation, both of management and workers; investment in time and training; and the quality organisation that is set up to implement the plan.

1. *Motivation*: The first step in active quality management is for every manager to take the quality plan seriously. Frank Caplan, an American specialist in quality systems, tells the story of the introduction of Zero Defect schemes into many American companies in the 1960s. The theory called for employees to re-create the old-time craftsman approach in their work. In return management were supposed to eliminate the negative conditions outside the workers' control. In some cases the effects were substantial, but in most the initial enthusiasm quickly died away when management failed to attack the root causes of quality problems. As Caplan put it, 'You see, it got to be a little tough for management to do this because it turned out that management has about 85 per cent of the problems to solve. They had leaped into this program on the assumption that the employees were to blame for all of the problems, rather than for 15 per cent of them.' For a quality assurance programme to work, it has to be driven from the very top of the company.

For management this means a long-term commitment. To play a good game of golf requires hard regular practice, and the better you are the more you need to work to improve. It's the same with quality management. In the beginning quality levels should improve by leaps and bounds. Then the improvement will slow down, and only hard pounding from management will even sustain that improvement. While there is a management 'drive' on quality, everyone's attention is focused on it. After a while the pressure of attention tends to weaken, and the old 'Sure, it'll do' attitudes creep back. At least that's what will happen if a fundamental change in management attitude has not occurred. All that will be left of the quality drive will be a few ill-understood extra bureaucratic procedures added to the production cycle.

Not only must management take quality seriously but also they must be seen to do so. The quality team should generally report to the managing director, not the production director. If the management team are seen to ask for the profit and productivity figures first and then the quality indicators only as an afterthought, supervisors and junior managers will quickly get the message. The quality idea must be constantly preached, but managers must be seen to take the hard decisions by rejecting substandard products and investing in quality disciplines that will convince the workforce that they are serious.

2. *Investment*: The second step is a large investment in time and training. Staff time has to be taken up in defining the key quality elements, and in analysing the processes that produce them. The new standards have to be evolved and written down, then attained and enforced — all of this takes key staff away from their daily work. Very often training programmes have to be set up to ensure that the new standards can be maintained. Although this investment in time and

money will be repaid, companies do have to start things by spending money. To engender confidence in this expenditure, senior management should select a bell-wether project. Named after the leading sheep with the bell round its neck, this is a test item to which quality disciplines can be applied, and the results monitored. A typical example would be the control of scrap on a line. If the cost of this can be discovered, and the causes isolated, the savings can be very considerable. Success in this project enables management to invest in further quality assurance planning with confidence. As the Chinese sage put it, the journey of a thousand miles starts with a single step.

3. *Quality organisation*: The final element in active quality management is the quality organisation. The style and design of this will vary from company to company, but in every case its task will be the setting and monitoring of quality standards. This first of all involves the quality management staff, who prepare and put into action the quality studies and analysis, and all the other aspects of quality planning. Critical to this is the selection of key quality indicators for top management reports, which should consist of a summary of results on a series of control elements. Typical subjects would be:
— customer complaints (by product, number and type)
— quality costs
— interim results of annual improvement programmes
— summaries of inspection reports.

The second element of the quality organisation is the inspection and testing staff, who are responsible for day to day control to check that the quality plans are put into effect on the shop floor. Quality costs are discussed in more detail in Chapter 25.

Developing a quality organisation

Begin with the creation of a quality project team, under the control of a senior executive. This could well be the team that combined to identify the relative quality profile of the projects at the very beginning of the process. This quality team will have as its first inputs the quality profiles and the target ideal quality system outline. This will probably be based on the list above, but it will undoubtedly be developed and modified as the introduction of the quality system goes on. The brief of the quality team will be to develop a detailed quality system for the company, to set periodic targets and measures of achievement, and to make a regular progress report to the chief executive.

The second stage will be to develop a training and research programme. This should start with the quality team, to enable them to bring the established techniques and vocabulary of quality systems into use. The course should concentrate on systems behaviour, motivation, diagnostic tools such as frequency diagrams, 80/20 analysis, statistics, etc., and relevant technical aspects. These techniques and others are described in Chapters 20 and 21. At this stage the aim is to give the quality team the ability and confidence to develop a working quality system. Later they will be expected to pass on some of their learning in the form of familiarisation talks to other workers.

The research aspect involves investigating the existing systems in the factory, with the cooperation of supervisors and workers. Existing quality standards, documentation, personnel relations, observation of machine capabilities, supplier assessment techniques, etc. should be looked at with a view to gathering information to enable the problem solving stage to be effective. Nothing alienates shop-floor assistance quicker than a feeling that the quality team simply doesn't have an adequate understanding of the problem.

The third stage is to examine a particular quality problem — perhaps one of the product attributes that the group noted as being inferior to the market average. The problem should be readily analysable, solvable and worth solving. It is important that the bell-wether problem should be successfully tackled. Using the Ishikawa 'fishbone' diagram, and other problem analysing techniques, with inputs from supervisors and workers involved (for more detail on how this technique works, see Chapter 20 on problem solving), the team should tease out the causes of the quality defect.

It might be something that can be remedied easily, such as the recalibration of a machine, or something more fundamental, such as motivating workers who can perform difficult challenging tasks faultlessly, but not dull routine ones. It might be that a similar quality problem occurs in more than one product, traceable to a common cause. As many as 65 per cent of defects in the Irish marketplace can be traced to the quality of bought-in raw materials. Obviously this is a major problem for the purchasing section, and for a company suffering from this kind of problem it might be appropriate to impose a total supplier assessment system immediately. Purchasing and supplier assessment is described in Chapter 4.

The example above illustrates how solutions can uncover new problems. A revised supplier assessment scheme implies proper checks on incoming material which requires staff trained to use statistics, and alert to the quality requirements. Inspection of suppliers' plants and discussions of the new standards will take considerable management time. But by solving problems one by one, always with the development of the total quality system in mind, the quality team can gradually work towards a total system.

This step-by-step approach has three advantages over the radical once-for-all introduction of a new system.

1. *Integrated change*: Since companies are changing all the time, it enables the quality system to be introduced in parallel with the change. The quality system then becomes organically part of the company, and not a separately imposed system.

2. *Involvement*: It is important that the benefits of a quality system be perceived by everyone, and that everyone in the firm be gradually drawn into a high quality attitude. This is more likely to occur through a series of small but impressive victories than by attempting one major triumph.

3. *Walk before you run*: It enables the quality team to experiment in small ways with new ideas, and to make the inevitable mistakes without jeopardising the whole programme.

Conclusion

Senior managers are responsible for the strategy of the company. It is their decisions that shape future growth over the long term. Research shows that a high relative product quality is a very significant cause of high long-term profits and return on investment. The impact of quality costs is another reason to implement a quality plan. Finally the forthcoming legislation changing the liability of manufacturers for injuries and damage caused by their products will make it essential for prudent manufacturers to set up documentation and testing systems that will stand up in court.

The most important thing senior managers do in implementing and sustaining a quality plan is to want it to happen. Almost everything else can be done by enthusiastic staff lower down the line. But if senior management don't show hands-on leadership, the plan will fail. They must demonstrate that the plan is important, both to the company and to them personally.

In order to establish and sustain this enthusiasm, management have to spend money on training. Good training well planned and given to the right people will always pay for itself. Simply sending Seamus on a course because he can be spared does nothing except give Seamus a holiday. Management also have to be prepared to let executives spend time developing the quality plan. Nothing really useful can be simply bought off the shelf. Every idea and technique contained in courses, films, in this book and the many others on the subject must be considered in the light of the individual company's needs.

One tried and tested technique is to pilot the quality plan approach on a bellwether project. If that is well chosen, the savings and improvements will rapidly repay the time and training expenditure, and senior managers will have the confidence to make the plan more general.

Chapter 3

Product design

DESIGN is the beginning of the industrial process. Designers develop the product from the start, specify exactly how it is to be made, and subsequently initiate or effect improvements. Every year thousands of new products are introduced to the marketplace, most of which fail. In the food industry alone 900+ new food products were launched in 1984. The rewards of a successful new product launch are high, but so are costs. This is partly because existing manufacturers deliberately try to create a high cost of entry into the market by extensive advertising and if possible by tying retail outlets to exclusive contracts. But even without these barriers, the process of new-product initiation is expensive. There are seven basic processes:
— concept
— feasibility and preliminary market test
— detailed design
— manufacture of prototype
— pre-production demonstration and testing
— design changes
— production and launch.

And that's all before a single customer uses the product! Once the product gets into the marketplace, the concept of *life-cycle product planning* comes into play. It is no longer enough simply to put a product into the marketplace, sell it and plead *caveat emptor* ('let the buyer beware') after that. The designer has to envisage the life-cycle costs of the product (costs of maintenance, repair and servicing) and the environments in which it is to be used.

Product design is the result of a series of compromises between the conflicting demands of customers, manufacturers, the law and the costs of things. The customer wants a product that is attractive, useful, safe, reliable, maintainable and cheap. The manufacturer wants a product that the company can easily make, preferably with existing equipment from easily obtainable raw materials, and one that the company can distribute and sell, preferably through existing channels. The law demands that the product be 'of merchantable quality', and in accordance with any relevant product standards. The designer has to tie all these demands together and create a product that can be made for less than the customers will be prepared to pay for it.

The aim of quality assurance in design is to ensure that products will perform

Product design

Good design is the beginning of quality *Kilkenny Design Workshops*

as required. They must also be safe and work with satisfactory reliability throughout their lives. The forthcoming legislation on product liability will make it necessary for design documentation to be complete, retrievable and accurate.

The qualities of design

When we think of design, we usually think first of all about looks. In fact appearance is the one design element on which the quality disciplines have little to say. Quality is more concerned with what the product does than how it looks. Five other design qualities are much more important: usability, safety, reliability, maintainability and traceability.

1. *Usability*: The basic requirement is for the product to do what it is designed to do, in the conditions that it is likely to meet. This is not always as easy to ensure as might be thought. Take perishable food products such as cakes or drinks — the designer has to take into account the length of the distribution chain and the time that the product is likely to be kept in warehouses, store-rooms and on dusty shelves. The designer has also to ensure that the product is usable by a diversity of people. The average height of Irishmen may be 69 inches (175 cm); but only a small minority of them are exactly that height. A product designed for that height may well be quite useless for everyone else (particularly those of the population who aren't men). Similarly, large minorities of Irish people are left-handed, wear glasses, are physically handicapped in some way, or are stout and short of breath. Product ranges that ignore these facts sin against usability.

Product instructions are a crucial aspect of usability. Unfortunately most people are reluctant to read instructions. Research in England discovered that 66 per cent of people fail to read instructions relating to food products (e.g. custard powders), 57 per cent don't read those relating to tools (e.g. shears), and 47 per cent don't read instructions relating to household products such as shampoos. This may be foolish, but since it happens, designers have to take it into account. The whole product includes the instructions and the wrapping.

2. *Safety*: No product can be absolutely safe. Life being as it is, there will always be some aspect of risk. The problem is to discover the level that is acceptable practically to the consumer and financially to the manufacturer. When serious injury or even death is a possibility, however, different criteria are called for. For instance products for children have to be especially carefully scrutinised. Serious injuries have resulted from babies swallowing detached bits of toys, or licking lead-contaminated paint.

Safety analysis starts by distinguishing the frequency of the hazard and the severity. Severity is usually divided into four classes:
— negligible: will not result in injury or product damage;
— marginal: can be counteracted or controlled without injury to people or major product damage;
— critical: will cause damage to people or major product damage, or will require major corrective action to prevent injury or damage;
— catastrophic: will cause death, severe injury or product loss.

In assessing the likelihood of damage, the various ways in which the product might fail should be considered both separately and together. The kinds of questions the designer should ask are: What is the likelihood that a failure or defect in any component will cause damage? What severity of damage is likely and possible (these are not the same)? How far will combinations of defects increase the likelihood and the damage? The most serious accidents are nearly always caused by a combination of circumstances.

Once this form of hazard analysis has been done, the designer has to reduce the level of hazard, component by component. There are various techniques that can be used. The hazard can sometimes be minimised or eliminated by using intrinsically safer materials, such as plastics rather than metal, water rather than electricity. Isolation systems and circuit breakers can shut down the system or part of it if a hazard is detected, on the same principle as the fuses in a domestic system. Some isolation systems enclose the potential hazards behind doors or closures. Finally, the effects of hazard can be reduced by monitors and warning systems that alert operators to potential dangers. If none of these techniques are adequate, the hazardous system will have to be controlled in isolation.

3. *Reliability*: From the customer's point of view, a reliable product is one that works well, and goes on working throughout its life, with a minimum of attention. Anything less is unreliable. From the manufacturer's point of view, reliability is defined statistically, as 'the probability that an item will perform a required function without failure under stated conditions for a stated period of time'. This is an average, since some products from the same batch will last longer than others for all sorts of random reasons. The proportion that have failed at any time after manufacture can be plotted in a cumulative graph. Typical measurements of failure rates include:
— Mean Time Between Failures (MTBF), or the average time between one breakdown and another of a repairable product;
— Mean Time To Failure (MTTF), or the average time a non-repairable product (such as a light bulb) goes before breaking down; also used for the period before the first breakdown of a repairable product, which may be referred to as Mean Time to First Failure (MTFF).

Failure rates are often pictured as following a 'bathtub' curve over time. In the initial stages they are usually very high, though decreasing, as the causes of non-conforming products are discovered. During the normal lifetime, the failure rate is envisaged as low and constant, until such time as the products reach the limits of their life, when the failure rate increases rapidly. Plotting these rates against time produces the characteristic bathtub shape. This concept is not statistically manipulable, however, because it is a model combining three separate stages (see diagram). A great deal of statistical work has been done to identify mathematical models for failure rates. The simplest model to use is the *exponential distribution*, which assumes that the average number of failures per thousand will be constant over the life of the product. This kind of distribution has been found to be suitable for rare but regularly distributed events, such as misprints in a book, or the chances of being hit by bird droppings.

Unfortunately product failures are not usually of this type. For instance, in many cases the failure will be a result of the sudden breakdown of one item. The distribution of failure rates will therefore be dependent on that item. This is more analogous to such events as sudden floods, and the breaking strength of rope, etc. The suitable mathematical model to use is called the *extreme value*

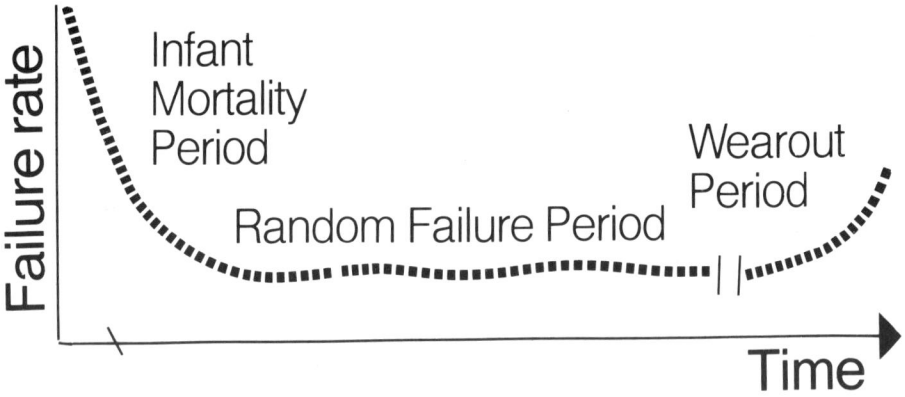

The 'bathtub' shape of failure rates over time

distribution. The *Weibull distribution* is often used in reliability analysis situations where non-constant hazard functions apply. This is typical of many products, which after time simply begin to wear out. It has also been used in analysing droplet sizes in sprays, sales distributions and a number of electrical and mechanical reliability functions. Yet another model, the *lognormal distribution*, is used where the effects of the initial fault accumulate, such as a crack in a part, or bacteriological attack. In each of these cases, the reliability engineer has to analyse the way in which failures occur before applying the appropriate model.

The study of unreliability, or failure rates, started seriously in the 1950s, when American engineers began to be worried about the failure rates of complex electronic devices. As electronics developed, devices became much smaller, which meant that they also became increasingly complex. Failure rates therefore increased, because the likelihood of failure is cumulative. Given an assembly made up of two components, with a reliability of 0.95 and 0.90 respectively, the reliability of the whole will be 0.95 × 0.90 = 0.855. Clearly, for a very large assembly, such as a ballistic missile with 40,000 parts or more, failure rates were likely to be high. The US Department of Defense and a group of electronic engineers set up an advisory group in 1952 which recommended a much greater degree of testing of individual components, monitored by statistical analysis. This quickly became standard practice. The current British Standard on Reliability of Systems, Components and Equipment (BS 5760) was issued in 1982.

The reliability of a product is strongly affected by design decisions. By the time the product reaches production, it is often too late to do anything about fundamental defects. It is essential therefore that the design process includes a sophisticated reliability analysis programme. Various techniques are used to assess the likely reliability of the manufactured product. These include:

Product design

— *Parts, materials and process analysis*, including established and novel features analysis. As design progresses, the reliability implications of each part should be considered, bearing in mind the stress levels likely to be encountered. As far as possible, parts and sub-assemblies whose performance is known should be selected. Features or parts being used for the first time should be subject to special analysis and/or testing.

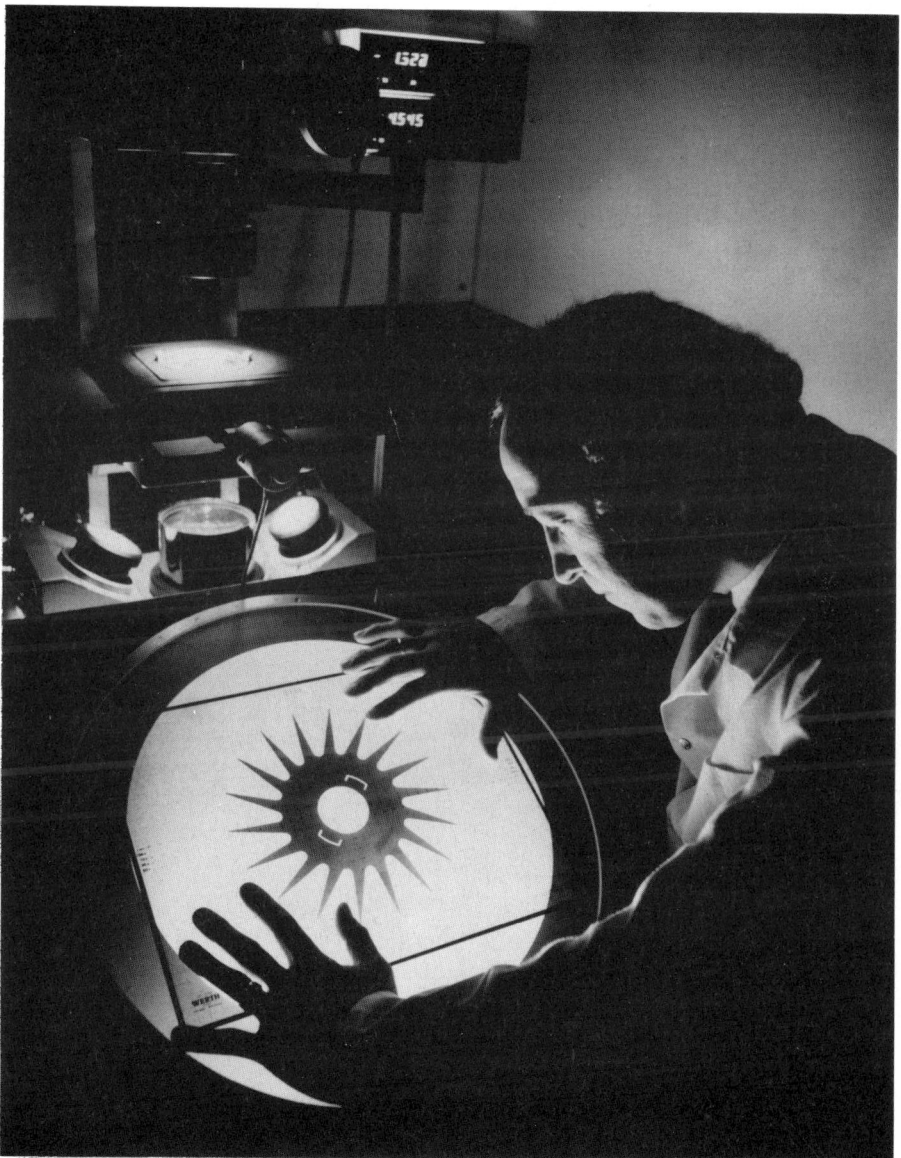

Checking design specifications   *Nypro*

— *Failure mode and effect analysis*. This is a technique for analysing potential failures, and their relative importance, or criticality. It is described in detail in BS 5760 and the US military standard MIL STD 1629 (Procedures for Performing a Failure Mode, Effects and Criticality Analysis). FMEA can also be used as a problem solving technique. The various stages in the analysis are as follows:
1. Identify the part or process and describe its function.
2. Briefly indicate the part's function.
3. List all the possible ways in which it might fail.
4. Describe the effects of each failure mode on the system as a whole.
5. List all the possible causes of the failures detailed under (3).
6. Estimate the probability of each particular kind of failure occurring (1 for low probability, 10 for virtual certainty).
7. Rank the severity of the failure in terms of its effects on the whole (1 is not serious, 10 is total failure of the product, or a safety hazard).
8. Rank the difficulty of detecting the failure before it reaches the end-user (1 for easy, 10 for virtually impossible).
9. Calculate the risk number for each failure mode by multiplying the figures estimated under (6), (7) and (8). This gives a *risk priority number*; the higher the number, the more serious the failure mode. From this list a critical items summary can be drawn up.
10. Give a brief description of the corrective action required against each failure mode, if possible.

— *Stress and worst case analysis*: This is an analysis to ensure that the cumulative effects of input conditions, loading and parts tolerances will not cause the whole system to fail. Its main purpose is to ensure that in all conceivable cases the safety margins specified are adequate.

— *Redundancy analysis*: For critical parts, it may be necessary to provide alternative means of performing the same job. This is called redundancy. The type and extent of redundancy will depend on the effects of failure of the part. If the two parts work in tandem, the stress on each can be considerably reduced. On the other hand, maintenance is increased, as is complexity, which is usually bad news for reliability.

— *Performance testing*: Once the design is off the drawing board and in prototype form, performance testing can begin. Parts and sub-assemblies can often be tested separately, and particular attention should be given to those which have already been identified as critical. Performance tests of the product as a whole are often not long enough to give much statistical information. As a result accelerated tests may be applied, in which parts or assemblies are subjected to more than normal use stresses. This can be done by constant high strain, or by a step process, where the pressure is built up gradually. Either method speeds up ageing and degradation, but because of the abnormal conditions, can provide irrelevant information. Two days in the Arctic may be no adequate indication of how one would survive a fortnight in the Scottish highlands.

The reliability of the machine cannot be considered separately from the human element. The operator's function, his or her information needs in normal and emergency conditions, the way in which the information is provided, the response required from the operator and the time allowed for that response all need to be considered as part of the overall reliability programme.

4. *Maintainability*: As ordinary industrial and consumer products become more complex, the cost of their maintenance during the whole product life comes to exceed the cost of original purchase. The designer should always therefore consider the design from the point of view of the repairer. For instance, components doing the same job should be grouped together or identified by common colour. They should be standardised, and the layout should be simple and accessible. Test features should be incorporated, so part failure can be quickly and accurately diagnosed. Lifting hooks or holes should be standard, and heavy parts below light ones. Castors or easy moving systems should be supplied for heavy equipment. Connectors should be easy to attach and disconnect, and clearly labelled. Servicing and lubrication points should be easily accessible.

Various general issues arise in considering maintenance policy that companies need to consider:
— Should the design maximise reliability or repairability?
— Should the company use standard modular parts, which require more design effort to incorporate but are easy to repair, or special non-standard ones?
— For some products or parts, the cost of repair exceeds the cost of replacement: how far should the company go along the throwaway route?
— Should test equipment be built in, despite the increases in design time, cost and bulk?
— Should the assembly be highly designed as foolproof for non-skilled operators, or more general for skilled operators only?

5. *Traceability*: The law of strict product liability in effect imposes an additional design requirement on manufacturers, that of traceability. If a product injures a customer, it will be necessary to trace exactly where and when it was made, which raw materials went into it and so on. Every part should be traceable back to the original batch records. But apart from legal requirements, traceability is essential to any reliability study. If the product's life history cannot be followed, no statistical analysis can be done as to how the failure rates for, say, small design improvements match up to previous failure rates.

Conclusion

When the Book of Kells was enscribed, Irish design led the Western world. Since then we have lost ground. Design has long been recognised as an essential part of the business of creating an industrial society. In 1962 a report commissioned by Córas Tráchtála from a group of Scandinavian experts criticised the lack of design awareness in Ireland, but saw this as a great opportunity to create a unique contribution to European culture, unhindered by the past as other countries were. In 1965 the Kilkenny Design Workshops were set up, since when they have been producing traditional and new industrial and craft design. As well as the familiar products, such as cloth (tweeds, curtain fabrics, etc.) and pottery, the Workshops have created industrial designs for products as diverse as glassware, box files for offices, stoves, a twelve-ton trailer, electronic equipment and packaging for tennis balls.

Good design is not just a matter of good looks. The interlocking requirements of safety, usability, reliability, maintainability and traceability place great demands on the designer before looks are even considered.

Chapter 4

The buying function

As industry develops, companies become increasingly specialised. A sophisticated product might have many hundreds of components, most of which are made by someone other than the final assembler. In Britain, bought-in elements constitute 70 per cent of the materials costs of electric cable, 60 per cent of chemicals, 65 per cent of motor vehicles, and 50 per cent of mechanical engineering products. It is easy to see why up to two-thirds of defects in products stem from defects in bought-in elements.

Purchasing activity has been at the heart of the development of quality control standards. Major purchasing bodies, such as NASA in America and the Ministry of Defence in Britain, who spend hundreds of millions of pounds on procurement, have led the way. General quality assurance standards such as IS 300 and BS 5750 have been developed to reassure purchasers that objective quality standards are being met. The standards authority certifies that specific suppliers have achieved a degree of quality excellence.

Sophisticated purchasing organisations can have a dramatic effect on the quality consciousness of companies that supply them. Until recently, for instance, Irish printers employed no separate quality staff as such. Companies generally relied simply on the quality judgement of skilled machine operators. In the marketplace speed of delivery and price were seen as critical. Then, in 1984, several large computer software firms began to have their computer manuals printed here. This was the first time that most of the print companies had been exposed to supplier assessment techniques, inspection sampling plans and the other techniques of modern purchasing. Because these companies had a lot of money to spend, demanded high standards, and knew the quality disciplines by which they could be achieved, most of the big Irish printing companies now have quality managers, quality control staff, and are developing quality systems suitable to the industry.

There are two main aspects of purchasing policy: *supplier assessment*, or the techniques for choosing suppliers, and *supplier relations*, or the techniques for ensuring that suppliers keep up the quality standards required. In American usage, the word 'vendor' is often used instead of 'supplier'.

Supplier assessment

1. *Component needs*: Bought-in components can be divided into critical and non-

Quality in Practice

critical sets. Critical components can be divided into two further groups. There are those components whose quality critically affects the quality of the whole product; they impinge directly on the key quality elements. Obviously if these are bought in, the degree of control that must be exercised is very great. The second group of critical items comprises those that are not so important in the quality picture, but are vital to the production process. The non-supply of a few trivial items can stop a production line. Finally, an item may have a critical effect on the overall costing of the product. Non-critical items are those that neither significantly affect the final quality (though of course they must achieve a minimum standard) nor are likely to hold up the production line.

There are three factors in every buying decision:
— *quality*,
— *price*
— *delivery*.

The minimum or maximum values of each factor will be specified by the user department. The quality characteristics will be defined by design, the price levels by costs and the delivery schedule by production. For each of the critical and non-critical groups of components, a different importance will be given to these factors. For components critical to the production process, the order of importance might be delivery, price, quality; for non-critical items, it is likely to be price, quality, delivery; for quality critical items, quality, delivery, price. The purchasing function's job is to juggle the competing pressures from the user departments (also called the *buying influences*) to produce the best mix of the factors.

Exhibit: Raw material control at FMC International, Cork
Outline statement: FMC will select suppliers of raw materials based on a vendor rating system. The four areas in which evaluation is carried out are quality, price, performance and facility capacity. All firms wishing to become suppliers are asked to meet the terms of the following procedure.
Purpose: The objective of the rating system is to evaluate fully how well the vendor's quality programme operates and how closely it conforms to FMC requirements.
Notes: The vendor rating system will be based on a table 0-5 as follows:
5 very good
4 good
3 average
2 poor
1 very poor
0 negative
This evaluation is best carried out by a qualified team from quality assurance and purchasing departments.
Procedure:
1. Obtain written specification from the potential supplier of the material to be supplied. Check that it meets FMC requirements.
2. If satisfactory, arrange a plant visit and evaluation.
3. Analyse a sample from the plant. Compare with supplier's certificate of analysis.

4. The manufacturing capacity must rate 4 or more, and the sample must conform to standards.
5. If the above is satisfactory, grant initial approval. Hold over the first three lots supplied for full analysis on arrival. If the supplies comply, then add the firm to the list of approved suppliers.
6. If supplies fail to meet specifications, then the technical manager or chief chemist will decide on disposition. If the off-specification parameter does not directly affect product quality a form of waiver may be obtained. A judgement will then be made as to future shipments from that supplier.
7. Every two years send a Supplier Questionnaire to each approved supplier to ensure continuing adequate quality standards.
Source: FMC Policy and Procedure Manual (slightly condensed)

2. *Supplier policy*: A major decision in purchasing is the number of suppliers for each component. As always in purchasing, the decision will be affected by the degree of criticality of the piece. For a very critical piece, it will usually be desirable to have at least two potential sources. This will provide practical protection against non-supply because of strikes or fires and against a sudden falling-off of quality standards in the supplier's plant. By providing an element of competition it will also ensure that prices are kept reasonably low. If the purchasing company is operating Just-In-Time production (see Chapter 7), different considerations, such as proximity to the factory, may have to be introduced into the calculations.

The disadvantage of multiple source supply is the loss of the close relations that a single source can provide. As the purchaser and supplier get to know more about each other's business, significant practical advantages arise. For complex products, this knowledge of the purchaser's business can be more valuable than a 10 or 15 per cent difference in price. With the surety of close relations, machinery can be made interdependent, with advantages on both sides. Managerial and economic advantages can also accrue as the two companies work more and more closely together, for instance by enabling the supplier to produce standard components in non-peak times. A purchaser who knows that an order is vital to the survival of a supplier is of course in a position to insist on good service.

3. *Choosing qualified suppliers*: Once the strategic decisions have been made as to the purchasing framework, it remains to discover suppliers who fit. They must first of all be able to meet the product quality requirements, the delivery requirements and the price. After that, general commercial elements enter into the analysis, such as financial security, trustworthiness, etc.

The potential supplier's ability to meet the quality standards can be assessed in three ways:

— *By test production*: A small sample run is commissioned and tested. The standards of test would be rigorous, because the production conditions are rather special. A supplier doing a test like this can be expected to take more trouble than usual over the work in the hope of obtaining a large order. Standards have a way of slipping thereafter.

— *By similar products*: The ability to produce similar products to the complexity

and standard required is a clear indicator of capacity. If the practice in the industry allows, it would be highly desirable to get an endorsement from other customers.
— *By capability survey*: This is a special form of quality audit (see Chapter 17). The auditors are concerned specifically with the manufacturing capability of the company, but also with the management and motivational factors that are so important to consistent quality. A supplier survey can be taken at any time: it may be a survey of a potential new supplier, or of a supplier who has been giving problems, or a supplier whose systems have been reformed. In the end, any form of supplier assessment is subjective. One study of 151 supplier surveys in the US discovered that in half of the cases the long-run results contradicted the assessment prediction. Supplier surveys are also expensive. To overcome these problems, the national standards associations have set up national quality standards systems. If a company is registered as complying with IS 300, or BS 5750, or ANSI N 45.2, then the purchaser can be confident that the quality assurance systems at least will be satisfactory.

Exhibit: Vendor assessment at Digital Equipment, Galway
The first quality assessment a vendor will encounter will be the process (quality systems) audit. In this audit particular attention is paid to
(a) quality management, philosophy, policies and procedures
(b) procedures; adequacy and compliance to procedures
(c) process controls (control charts, sample inspection)
(d) change controls (documentation, specifications, product, process)
(e) calibrations (traceability and periodic schedules)
(f) MRB process
(g) purchased material control (qualified vendors list, purchasing, quality).
The audit is especially to check the quality of the closed loop systems in all aspects of the operation. Additionally, plant safety procedures are reviewed, particularly as they affect the potential of the vendor to continue supplying in the event of a disaster.

The next quality systems assessment generally brings the vendor to 'ship to stock' status. Much of the systems assessment is common to the qualified vendors list type of assessment; the main difference is in the definition of the characteristic accountability charts and the final outgoing quality assessment of the product, where mutual standards are agreed.

Supplier relations

Once a supplier has been chosen, it is in the purchaser's interest to develop a close working relationship. The delivery of the product shouldn't be seen as the end of things. Good suppliers are not so common that they can be picked up and dropped at will.

1. *Joint quality planning*: The purchaser's quality policy should be made very clear

to the supplier. The position the supplied item has in the production process is important. Very often the supplier will know something of a component's capabilities that the purchaser will not. If the supplier isn't told exactly how the item fits into the product, critical but obscure quality requirements may not be met, for instance as to melting point, strength under stress, etc. Copies of sampling plans and acceptable quality levels (for more information on AQLs see Chapter 22) used by the suppliers should be made available. As the relationship develops, the supplier should become involved in the planning of quality improvements and design changes. Systems for exchange of inspection reports and rapid exchange of information on non-conformance should be set up. Ideally the computer systems of the supplier and the purchaser should be compatible, so that test results can be transferred directly, with the data communicated via modems and the telephone system. Another very valuable part of this mutual relationship can be joint training of quality and production staff.

Exhibit: Supplier/purchaser relations
Ballyclough Co-operative Creamery has traditionally paid particular attention to customer needs. Since 1932 it has developed a special relationship with Rowntree-Macintosh. Ballyclough provides the factory, building, milk, steam and effluent plant facilities to the Rowntree-Macintosh company, and in return is guaranteed an outlet for its milk at a guaranteed price. In 1984 the Mallow plant produced enough chocolate crumb to fulfill Rowntree's entire Irish requirements and half their UK needs.

2. *Batch testing and approval*: Incoming stocks should go into the inspection bay (a quarantine area, from which nothing should be allowed out without being labelled as approved or rejected), accompanied by a purchase order and a materials received docket from the unloading bay. If the supplying company is expected to provide the results of its own tests, these should be included with the batch. The inspection bay should contain the quality manual specifications and testing standards and procedures. Inspection and test procedures and documentation standards are discussed in Chapters 22 and 23.

Inspection procedures have three functions. They should
— test the incoming products for conformance to specifications,
— set up batch records for monitoring the flow of that batch through the factory, and
— record test results for supplier performance evaluation.

3. *Continuous assessment of suppliers*: The relations between purchasers and suppliers are constantly shifting. Management personnel change, pricing policies are radically altered, customer policies take different tacks, new suppliers come on the scene. As a result companies must constantly monitor the performance of their suppliers.

The factors of quality, price and delivery have to be monitored separately, but also drawn together in some kind of composite index, so that a composite judgement can be made. The usual way is to give each factor a numeric weighting

Bacteriological testing of cream *Virginia Milk Products*

Cream being delivered to Bailey's through a closed system to reduce risk of contamination
R. & A. Bailey

The buying function

First-Sample Inspection Report

BRAUN

Number: 850930A
Page 1 of 2

Sample Description: ☒ New part ☐ Changed part/index ☐ New tools or fixtures ☐ Multiple tool ☐ Changed tool ☐ New supplier ☐ Raw part ☐ Semi-finished part ☐ Finished part

Reports are available: ☐ Material Testing ☐ Electrical Testing ☐ Reliability ☐ Functional Test

Description: Display Carton
Part Number: 4.559.033
Change Index: 'O' 16.8.85
Tool No:
Cavity No:
Supplier:
Buyer/Telephone: G. CLANCY / 329

Delivery Number: 14,000
Delivery Date:
Receiving Number: 016502
Receiving Date: 26.11.85
Sample Quantity: 5

✓ = in Tolerance
f = out of Tolerance, must be corrected
f✓ = out of Tolerance, however accepted

Item No.	Characteristic, Dimension or Specification	Supplier- Inspection results	Recipient — Verification of Supplier Inspection Results	Rating
1	Size:	L: 227 +1 B: 91 +1 H: 40 +1	228 92 41	✓ ✓ ✓
2	Paper Weight		404 gr/m²	✓
3	Print Content & artwork		Correct for product i.e GC 40 US Clicker Version	✓
4	Print & artwork quality		clear & legible, no smudging or misalignment	✓
5	Glued edge		secure & neat	✓
6	Folding		correct to drawing	✓
7	Colour		Grey	✓

NOTE: All dimensions outside tolerance must be corrected before next run.
Sample-Weight (gramm):

Inspected by: Brigid Doran
Date: 27.11.85
Responsible Supervisor:

To be completed by Supplier
Samples made from final production tools? ☐ yes ☐ no
Additional hand operations performed? ☐ yes ☐ no

We hereby certify the above inspection results to be correct and that our sample meets the Braun specification
Authorized Signature
Date

To be completed by Recipient
Decision: FREE
Authorized Sign. Qual. Control
Date: 27.11.85
Factory: CARLOW
Department: QA

Distribution List:
INCOMING INSPECTION
MATERIALS

which will enable the buyer to compare the performance of company A, whose price is low, but whose reject rate is relatively high, with company B, whose delivery is occasionally unreliable, but price and quality are about average. The technique is to assign values out of a hundred to each factor, judging by its importance to the company. The weighting will be different according to the importance of the factor. Thus the purchaser might assign 50 points out of 100 to quality (judged by percentage rejected), 30 points to cost (judged by ratio to lowest possible price) and 20 to delivery reliability (judged by percentage of delivery dates achieved).

Exhibit: Supplier assessment chart

Factor	Weight	Measure	Score	Points
Quality	50	% accepted	85%	50 × 85 = 42.5
Price	30	lowest/this	75%	30 × 75 = 22.5
Delivery	20	% on time	90%	20 × 90 = 18.0
Total points in assessment				83.0

This supplier scores 83 points, which can be compared with other suppliers' scores. The scheme depends heavily on factor weighting, but as long as the same factors are given to companies producing comparable qualities, the result provides a reasonable method of comparison.

If a supplier consistently achieves high quality levels, it is reasonable to relax the supervision. The most commonly used sampling plans (see Chapter 22) allow both for tightened (more rigorous) and reduced inspection levels. Thus if a supplier has successfully delivered a number of batches without problems, he may be put on reduced inspection, whereby a smaller sample of the incoming batch will be examined. This naturally saves time and money. On the other hand it increases the risk that nonconforming products will slip through. As soon as a batch is rejected, or there is a break in delivery patterns, normal sample levels should be reinstated. In one procedure, if two out of five consecutive batches have been rejected, the supplier is put on to tightened inspection level. This means that larger samples are taken from each batch, to increase the likelihood that defective products will be detected. The extreme level of confidence is reached in ship-to-stock suppliers, whose record has been so good that no inspection is required at all. The ship-to-stock system is the heart of Just-In-Time production systems, which are discussed in Chapter 7.

Conclusion

Because so much of any product is made of bought-in materials, the purchasing function is critical to good quality control. In order to ensure that suppliers continually provide a quality product, the company's purchasing policy must be prepared with care. The bought-in materials should be divided into critical and non-critical categories, and the policy in relation to single or multiple suppliers worked out.

Potential new suppliers must be carefully assessed. They must be aware of the quality requirements of the company, and the various inspection and test plans in operation. The closer a supplier company can work with a purchaser, the better for the quality of the ultimate product. The inspection plan laid down in the quality manual should be followed exactly, with newly delivered stock kept in quarantine until it has been tested against specifications. Only in extreme cases should any stock be allowed out of the quarantine area without testing. If production necessities do dictate that this must be done, the batch should be specially flagged and tightened inspection levels imposed thereafter.

The inspection process is not simply a gatekeeping process, letting in the good and keeping out the bad. It has two other functions. The first is to set up the batch record, so that if problems or unexpected benefits are discovered later in the production process, the supply can be traced to source. The second is to maintain the suppliers' continuous assessment record. This enables the company to keep up to date on the effects of the shifts in the various supplier companies. Every company should be monitored for quality levels, for price and for delivery achievement.

Once the stock has passed from the quarantine area after testing, it should be clearly labelled as such and correctly stored. It is important at this stage to ensure that storage conditions are suitable to the product. Food and drug products, for instance, should be very securely and hygienically stored. If the stock has a short shelf life, a rigorous system of control should be used to prevent out-of-time stock being released to production. In any case it is normal to rotate stock on a first-in, first-out basis.

Chapter 5

In-process control

THE quality of manufactured products depends firstly on the quality of design, secondly on the quality of the bought-in materials and thirdly on the manufacturing process itself. The relative importance of these factors varies according to the product, but it is always the manufacturing section, the factory floor, that acts as the melting-pot, where the product emerges from its constituent parts. Unfortunately products put on to the market too often fail to achieve the targets specified by the designer. These failures can be divided into groups:

— *Failures due to deterioration*: The parts of a motor wear out because of friction, colours fade in the sunlight, springs become less bouncy after use. Such problems, caused by deterioration, have to be considered in the design and purchasing rather than manufacturing stage, though in most cases factors such as the quality of finish, manufacturing technique and operator skill can make a difference.

— *Failures due to manufacturing imperfections*: These are the failures induced into the product by the operation in the factory. Some of them will be caused by human error, perhaps induced by boredom and lack of motivation, some due to design faults, as for example where the combination of sub-assemblies at the limits of their tolerance produces a fault, and some due to simple bad workmanship caused perhaps by lack of training.

— *Failures due to external factors*: Apart from the product itself, failures can be caused by the operating environment, especially if the product is exposed to extremes of stress or weather.

From the beginning of industrial manufacture, managements have grappled with the problem of producing good quality. At first quality control was left to operators and foremen. As factories became more complex, however, inspectors were appointed to oversee quality. Then it was realised that what looks like 100 per cent inspection is, for human and practical reasons, even with the best will in the world, usually more like 80 per cent in fact. What's more, there was no scientific way of describing the probabilities of getting a non-conforming product. As a result, in the 1920s and 1930s statistical quality control was developed. It was widely used for weapons and aircraft production during the Second World War. Quality control was still a production matter, however, until it was discovered that up to two-thirds of product faults are caused by bought-in materials. At the same time it was realised that however sophisticated manufacturing control is, no system can improve on a bad design or bad raw materials. Sophisticated buyers then began to demand overall quality assurance systems from supplier companies, using the elements of the quality disciplines described in this book.

The kind of quality system that is appropriate will depend very much on the system of manufacture. The first distinction is between a *job* shop and a *mass production* shop. The distinction is not hard and fast. In general a job shop caters for many different customers, and is capable of a wider variety of products and designs. A mass production line tends to be specific to a product range. Job shops themselves vary from those producing large complex pieces of equipment such as production machines or locomotives to those producing small scale relatively simple products such as fashion garments or printing materials. The key distinction between the two styles of manufacture lies in the proportion of effort devoted to planning or control of production.

Job shop production control

The typical job shop receives an order, executes it, delivers it, and may never hear from the customer again. In this circumstance there is no second chance if the quality is wrong. Therefore the company must devote a high proportion of its effort to planning production. There are four questions that must be answered.

1. *Can the factory fulfil the design specifications?* The job must be considered in detail before it is accepted, to ensure that the company can in fact produce those widgets to the tolerance required. Especially with a sophisticated customer, this may not be as obvious as it seems. If the factory is geared to a loosely defined 'industry standard', it may not be able to meet very tight tolerances in practice.

2. *Is the process running as planned?* The factory may be well able to meet the tolerances required, but is it in fact doing so? To answer this may require devising and setting up special quality checks for every job. Normally, however, some version of the ordinary process control chart will ensure that the process is within the planned limits (see Chapter 23 for more about these charts).

3. *Is the output satisfactory?* The natural variability of industrial processes will ensure that some of the output is not satisfactory. Most customers will allow for this by specifying that no more than x per cent of the product should be defective (this is called the 'acceptable quality level' or AQL). Defects are often divided into critical, major and minor, depending on their importance. The first few products off the line should be examined in detail against design specifications, and thereafter key specifications should be checked by sample from the suitable sampling plan (see Chapter 22 for more about sampling). And finally, once all is done and delivered, the last question:

Quality in Practice

4. *Could we have made it better in any way?* The systems approach, the quality approach, demands:
— no work without inspection records
— no records without analysis
— no analysis without development.

In other words, the whole quality system, as well as attempting to control present production, must be geared to improving future production. The work done after a job is delivered in analysing and discussing how it might have gone better will shorten the work to be done at the planning stage of the next job.

Mass production systems

Another way of classifying production systems, particularly mass production systems, is by the type of quality controls that can be exercised on them. Apart from service-type operations, which are dealt with in Chapter 10, there are four types.

1. *Intermediate check type*: This includes most machine work such as lathes, pressing machines, etc., where the output can be quickly and easily checked as soon as it is produced. The checking procedures are usually a combination of those outlined for the job shop above:
— first-off check against design specifications, applied every time the machine is restarted or a substantial change of operating conditions occurs
— operator checks as the process continues
— patrol inspector checks from time to time (usually specified as at least two visits to each production point per shift
— close-off check to ensure that the last few at the end of the run are correct, and that the tool doesn't require maintenance.

Processes in which this kind of checking system is appropriate include diecasting, sheet metal work, plastics, and any operation in which a simple attribute specification such as length can be checked.

2. *Final check type*: In some kinds of process, particularly food and beverage mixing, very little checking is possible before the product is complete. As long as the equipment is clean and the inputs right, there is nothing that can be done until the output is tested. Thus intermediate stage testing for some products in the process industries sector, or chemical and other processes is simply not possible. In many of these cases any testing that can be done is to destruction, so there is also no possibility of 100 per cent testing. To cope with this problem, the food and drug industries have built up an elaborate code called Good Manufacturing Practice (GMP), which is discussed in Chapter 6. In general, however, this style of manufacturing puts considerable emphasis on the production and process planning, the equipment and the system. This is on the assumption

In-process control

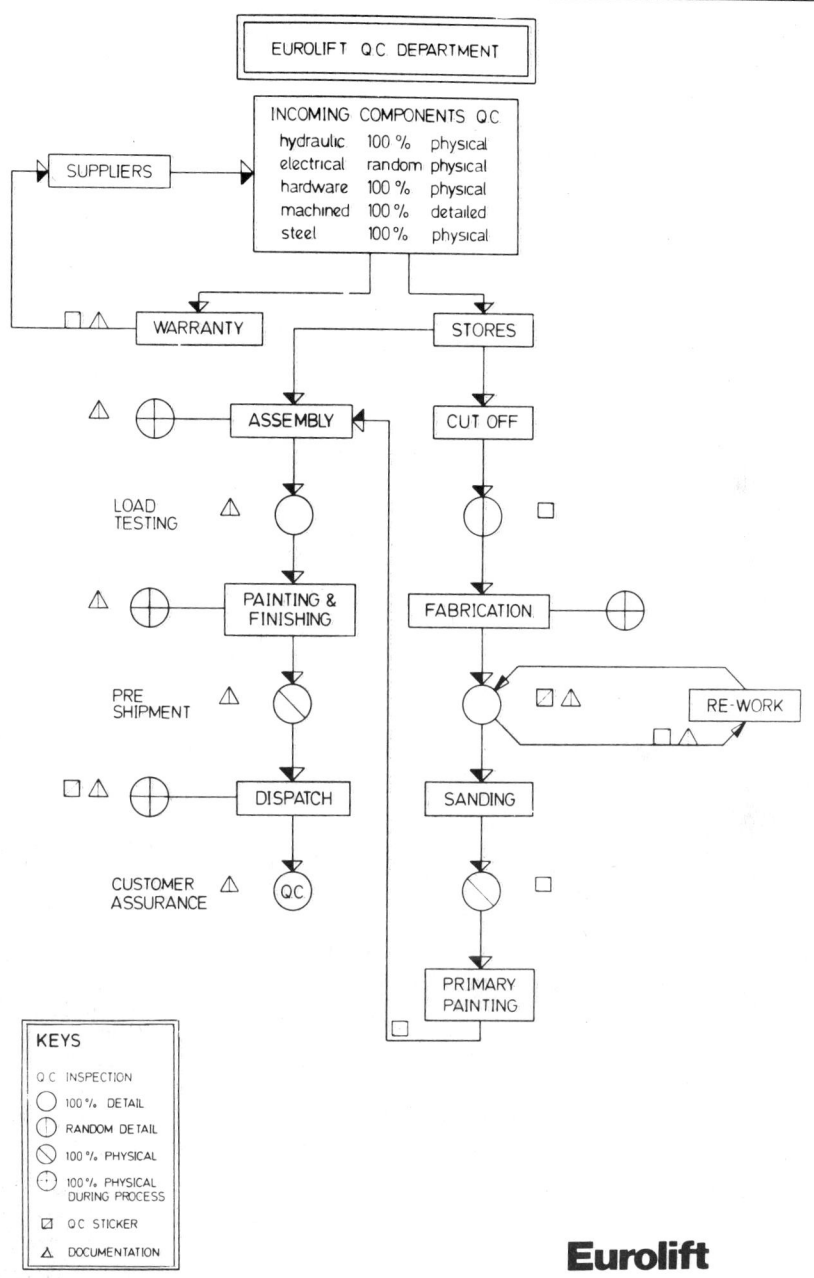

Chart showing location of process inspection points   *Eurolift*

that if the right ingredients are correctly handled, the right output must result. The first-off checks for each new startup are important, as are rigorously scientific sampling and inspection.

3. *Assembly systems*: This includes those cases where large or small components are fitted or fixed together. The quality control systems will be basically the same whether the system consists of assembling wooden dolls or making a rocket. The key steps in process control, once the desired standard has been specified, are:
— *Plan an assembly method*: This is the area of work that Taylor and the late nineteenth century proponents of 'scientific management' dissected so vigorously. Substantial savings can be made by designing the product and the assembly techniques to be efficient. This may involve, for instance, colour coding parts or designing pieces to fit only one way. Since the method or the order of assembly may be critical to the operation of the product, a method card, showing the steps and key points of the assembly, should be provided.
— *Materials flow*: Critical to the success of an assembly system is the way in which the substocks of materials at each assembly point are organised. One popular modern method is the very demanding Just-In-Time (JIT) method, whereby each assembly worker has two baskets for each input item. As one is emptied by use, it is instantly replaced by another, which is full. The amount in the basket is just enough to complete a cycle of work. The JIT system, which is usually worked right back to the suppliers (the ship-to-stock plan), is dealt with in detail in Chapter 7.
— *Inspection and checking*: Patrol inspection and rigorous testing of the end-product are important.

4. *Automated and computer-aided production*: Automated systems, and particularly computer-aided manufacture (CAM) systems present quite different quality assessment problems. The variability that humans introduce into operations is reduced to a minimum. At the same time, computers can be programmed to react to tiny variations in input conditions so as to optimise costs, output speeds and quality. There are, however, considerable problems in designing and programming computers for this kind of control work. It is much easier to design robots to undertake such formalised tasks as car underbody painting, inspecting medical tablets, or moving materials from one part of the factory to another. In terms of quality control, these systems have to be treated as if they were of the final check type. Once the process is designed and the computer programmed, the main concern should be testing the output.

Exhibit: Automated systems at Dairyland Ltd, Gorey, Co. Wexford
Automation must be considered as a way of improving productivity, but mainly of securing quality. Its most important feature is that it ensures that conditions set out for production are observed because it avoids the 'human mistakes' element.

The packaging lines (for Yoplait and other dairy products) are fully automatic. They form, fill and seal in one operation. Both product and packaging systems are totally enclosed and untouched by humans. All temperatures and settings are controlled by the microprocessor system. All the values have to be observed otherwise the plant will not function. The risk implicit in storing pre-formed packages is greatly reduced. A comprehensive and detailed quality control system can be linked to an automated plant, with a very wide range of data on the printouts recording times, temperatures, quantities, etc.

Operator control and its introduction

When managers of factories first started to investigate how they could improve productivity, they began by taking responsibility out of the hands of the workers. Machine operatives were to act as mere supplements to the machine itself, without any individual initiative or power. Each movement and sequence of movements was precisely prescribed. Unfortunately humans make rather inefficient machines, so much of the output was substandard. Systems of inspection and quality control were introduced to remove the non-conforming items. This was of course extremely expensive; some factories were spending 25 per cent of their time working and reworking defective products. At last someone had the brilliant idea of encouraging the operatives to make the product *Right First Time*. They realised that quality is the result of two activities:
— quality manufacture and
— quality control.

Inspection systems can only reject bad quality items — they cannot inject quality into the output. The only people who can do that are the people who actually make the product: the operators. If they are asked to make the product and to check their own work immediately it is done, then all sorts of benefits can occur. First of all operators suddenly become interested in the quality as well as the quantity of their output; secondly, problems are detected almost as soon as they arise; thirdly, the police-state atmosphere of the relations between production and quality staff can be significantly improved. The idea of *operator control* requires that managers change from the traditional view of workers as lazy, work-shy and motivated only by money, and treat them as people. As people, they have their own drive towards excellence. Managers can develop and use this to increase the motivation of the operators by allowing them some control over their own output. Of course operator control is not the answer to every problem. It requires
— clear, easily understood, non-subjective quality standards
— production machinery able to meet these standards
— measuring gauges and other instruments for operators
— training for operators in the use of measuring tools
— adequate time and good documentation to allow the operators to perform and record the tests

Quality in Practice

Operator control in action   *Chemoflon*

— check systems by patrol and final inspection.

With systems of operator control in place, companies have experienced cuts of 30 per cent in total scrap and rework charges, reducing inspection costs at the same time.

Operator control is a considerable change for a normal factory. It requires that the operator take responsibility for the output in a way that has not been required before. Naturally some workers will be concerned that their output may not meet the new standards, and their jobs will be at risk. Unions will be very alert to this possibility. The introduction of the system must therefore be discussed carefully with the unions and the workers generally, with the usual 'no redundancy' guarantee. The system requires mutual trust between management and workers. If that trust is not present, it will be extremely difficult to make changes. There are four basic steps towards the introduction of full scale operator control.

1. *Define objectives*: Management should think about the results they wish to achieve. Is the idea primarily to increase productivity, to increase average production quality, to improve morale, or what?

2. *Investigate resources*: How far is it practical to define adequate yet simple quality checks that can be done by the operators more or less where they work? Clearly sophisticated chemical work or very precise measurements would be unsuitable. Are the quality standards themselves sufficiently objective? Can the machines be expected to meet these standards? The new system will put operatives on their mettle, and perhaps on the defensive. There is no point in introducing operator control if the result is merely arguments about whether the machine or the operative has failed to meet standards. Are there resources and time available to train operatives in the new techniques?

3. *Analyse response*: The new system will change the relations of operatives, inspectors and managers. Is it possible to foresee how things will turn out? The introduction of such a plan will require careful selling to each of these three groups, and the response should be monitored.

4. *Plan introduction*: Start with a small self-contained part of the factory, and introduce a pilot scheme. Make sure that the workers are fully aware, through training and practice, of exactly what they are expected to do. Monitor the figures from that section, and compare them with the rest of the factory. Keep the section informed, and encourage them to air their problems, questions and complaints. Make a log of the kind of problems that crop up, so that when the scheme is introduced to the rest of the factory, they can be solved in advance. If the experiment is a success, move the system slowly into other sections. It is often better to move section by section rather than attempt to do everything at once.

Operator control techniques

Operator control requires that the operator ensures that his or her work complies with the specification. In order to fulfil this responsibility, certain requirements must be met by both management and the operator.

1. *The specification*: The most important of these is the specification of the standard. This should be in the form of simple instructions unambiguously explaining the dimensions or types of assessment that are required, and the rules for operating the various measuring devices. The specification should also clearly explain the inspection procedures (100 per cent testing or by sample, and if by sample, how the sample is to be picked and how often). It should also explain how the documentation is to be filled in. The accuracy of the specification is not the responsibility of the operator.

2. *The operator's responsibilities*: The operator must make the product as before, fulfil the testing specification, and also take a more general responsibility for the quality. If the goods are clearly defective in some way, it will be the operator's

Quality in Practice

 No. 2

PRODUCTION RECORD TETRA PAK

Machine No: Time at Start of Production: Time at End of Production: Reading of Pack Counter:

MACHINE CHECKS BEFORE PRODUCTION

Function	Range	Actual Pressure
Water Pressure	3.0 – 4.5 bar	
Air Pressure	6.0 – 7.0 bar	
Jaw Pressure	4.0 – 4.5 bar	
Melting Pressure	3.5 – 4.5 bar	
Flap Sealing Press	2.5 – 3.0 bar	
S.A. Pressure	0.2 bar	
L.S. Pressure	0.2 bar	
Lub. Oil Pressure	4.0 – 5.0 bar	
Cooling (Left)	2 /Min	
Cooling (Right)	2 /Min	

PAPER & STORE INFORMATION

	Time	P. Order No.	Reel No.	Time of Strip Splice
Start				
1st Splice				
2nd Splice				
3rd Splice				
4th Splice				
5th Splice				
6th Splice				
7th Splice				
8th Splice				

PACK CHECKS

	St.	20'	40'	Spl	20'	40'	Spl	20'	40'	Spl	20'	40'	Spl	20'	40'	Spl	20'	40'	Spl	20'	40'	Spl	20'	40'	Spl
Destruction per D.S.																									
Overlap Sealing																									
Weight Check																									
Flap Sealing Check																									

OIL LEVEL CHECKS

Central Lubrication		
Pressure Header	L	
	R	
Mist Lubrication		
Work Gearbox		
Index Unit		
Bevel Gearbox		

Date ....

Signature ....

An operator control chart   *Drogheda and Dundalk Dairies* responsibility to report this to the inspectors. Depending on the specific sampling plan, the operator will be responsible for ensuring that the first-off test is done by the inspectors, and that the ongoing control charts are completed. When the work is defective it is the operator's responsibility to inform the inspectors or the foreman immediately.

The bin system is one way for the operative to keep tested from untested output. As soon as the newly made widgets are taken from the machine, they go into bin 1. They are then tested, 100 per cent or by sample. If they pass the test, the result is recorded and the tested and approved output is put into bin 2. Any rejected output is put into bin 3. A variant on this procedure is to have a fourth bin into which only the tested sample goes, so that the inspector can check the testing procedure.

3. *Defect procedure*: Under operator control systems, defects are divided into four categories:

1. Operator discovered defects: operator liable
2. Operator discovered defects: operator not liable
3. Inspector discovered defects: operator liable
4. Inspector discovered defects: operator not liable.

If the operator discovers defects caused by his or her own workmanship, then as far as possible he or she should sort and rectify. Other faults should be brought to the attention of the inspector, as machine adjustments will probably be required. As a general principle operators should be paid for all output. This does not mean that they should always be paid at the same rate. For defects caused by the operator, some companies charge half the cost of scrap and rework if they were discovered and reported by the operator and all the cost if they were discovered by the inspector. The records of the system should allow production and quality management to monitor the average quality level of each operative, and so regulate the degree of inspection normally required.

Conclusion

Operator control is the most satisfactory method of controlling the manufacture of goods. Most modern quality campaigns, whether Zero Defects programmes or Right First Time programmes, boil down to the fact that it is impossible to inspect quality into a product. It has to be made correctly from the start. Quality assurance means quality in the making controlled by a quality plan that directs the design and manufacturing effort. A key part of this is the operator control plan.

Operator control, however, does not do away with the need for inspection, but is designed to add to the inspection procedures. It also acts as an early warning system. Instead of waiting until the patrol inspector gets round to checking the output, each operator can keep a constant monitoring eye on quality, thus avoiding the manufacture of large batches of non-conforming products.

Chapter 6

Good Manufacturing Practice

QUALITY is important for all product types, but for two it is vital. These two are food and drugs, including sweets, drinks and cosmetics. Because of their importance, a special set of regulations called *Good Manufacturing Practice* (GMP) has evolved to control their manufacture. These regulations were first issued by the American Food and Drugs Administration in 1963, in the aftermath of the Thalidomide tragedy. They have since been updated, and adopted by many companies as standards. GMP standards represent the very highest level of quality aspiration in manufacturing.

The events surrounding the sale of Thalidomide revealed how vulnerable the drug-consuming public were to error and greed among the drug companies. The drug was marketed as a sedative. It was sold in Britain, despite the fact that questions had been raised about it in Germany, and it was being withheld in the United States. The appropriate department of the British government had sanctioned its sale on prescription. Unfortunately the drug had a tragic side-effect. Taken in the early stages of pregnancy it greatly increased the risk of babies being born with stunted arms and legs. Several hundred of these children were born in England alone. Considerable legal effort was required before the company that sold the drug admitted legal responsibility. Several other examples of exploitation by the drug companies of their dominating position in the 1960s and 1970s have led to much stricter scrutiny by governments of the whole market. At the same time, regulatory agencies began to realise that food products were also potentially hazardous.

As more and more processed foods are being consumed, the public are becoming as vulnerable to injury from food as from drugs. The EEC has listed over 600 permitted additives that may be mixed with food products. These are designed to colour food, preserve it, sweeten it, prevent oxidation, emulsify and stabilise, make it shiny, make it bulky, enhance flavour and so on. One additive, polyphosphates (E450), enables manufacturers to add water to meat — as one advertiser put it in a trade magazine, 'why sell meat when you can sell water?' Some food additives can harm vulnerable sections of the population such as asthmatics, diabetics and children. As well as these permitted additives (which are less than 10 per cent of those available) there are some 5,000 flavourings available, which are not regulated.

Apart from the inclusion of additives, which make processed foods manufacture more and more a subset of the pharmaceutical industry, there is also straightforward carelessness in manufacture and packaging. In 1981 the Dublin region public analyst reported discovering in various products with sealed containers pieces of glass, cockroaches, beetles, bluebottle eggs, wasps (in jam — presumably the maker had used real fruit!), penicillin mould, bits of a grasshopper, earwigs, wire, spider beetles, and anti-freeze in lemonade. In 1985 anti-freeze components were added as a sweetener to wine in Austria, and in 1985/6 Italian table wines were chemically adulterated. At least twenty people were known to have died as a result of the latter incident.

The Austrian wine had in fact been tested and passed by public analysts. EEC wine manufacturers are permitted to add up to twenty different chemical additives to wine, and a further twelve fining (clarifying) agents. Because of the huge number of possible substances in wine, the Austrian wine was tested only for certain chemical substances. Anti-freeze, not unreasonably, wasn't one. As a result the wine poisoned several hundred people. The same limitation applies to the testing of drugs. Because every remotely possible substance in a drug might require a different test, it is usually practical to test only for what is presumed to be there. This technique, called *presumptive testing*, means that the product is unlikely to be tested for substances introduced as a result of wrong labelling or mixing. Drugs are carefully tested for what is expected to be there, and perhaps some general substances such as those caused by decomposition. They cannot be tested item by item for what isn't supposed to be there.

Another cause of potential problems in food and drug testing is that many tests are totally destructive of the product itself. The only way to check whether a beetle larva has accidentally got into a cake is to cut the cake open. But then it can't be sold. The same applies to the more sophisticated tests that are applied to drugs. To test for active components, they must be released from the main part of the tablet, the inactive components. In either case, 100 per cent testing is impossible. As a result test samples have to be drawn. Carefully drawn samples can be expected *in the long run* to represent accurately the population from which they are drawn. There is always unfortunately a small possibility that the sample may be totally unrepresentative.

Sampling is discussed in more detail in Chapter 22, but the point can be understood by analogy. Suppose thirteen cards are dealt from a pack; they represent a sample of the whole, and would be expected to contain a mix of suits and a mix of court cards and number cards, just as the main pack does. There is however a tiny chance (actually 158,753,389,899 to 1) that the thirteen cards picked from a normal pack will be all one suit. If the sample turned out like this, the smallness of the chance would tempt any unbiased observer to investigate the possibility that the pack contained nothing but spades (or whatever) and act accordingly. One would be less likely to treat the result as a dramatically unrepresentative sample from a normal pack. In a card game this mistake is unlikely to be serious. A defective batch of drugs, if put on to the market as a result

Good Manufacturing Practice

of the operation of remote chance in this way, could have fatal results. Of course probabilities work both ways. There is a similar risk that any batch will be rejected by the operation of probabilities despite being perfectly satisfactory. In this case it is the company's money that is lost, rather than the patients' lives.

The code of Good Manufacturing Practice is designed to create as much safety for the consuming public as is possible. It is not of course the only safeguard in this respect. Indeed laws have covered the adulteration of food since the Middle Ages. Present legislation includes the Sale of Food and Drugs Acts, 1875–1936, the Milk and Dairies Acts, 1936–56, the Health Acts, 1947–70 and the Food Standards Act, 1974. The GMP regulations themselves cover ten substantive areas:

1. Organisation and personnel
2. Buildings and facilities
3. Equipment
4. Control of components and containers
5. Production and process controls
6. Packaging and label control
7. Warehousing and distribution
8. Laboratory controls
9. Records and reports
10. Other regulations

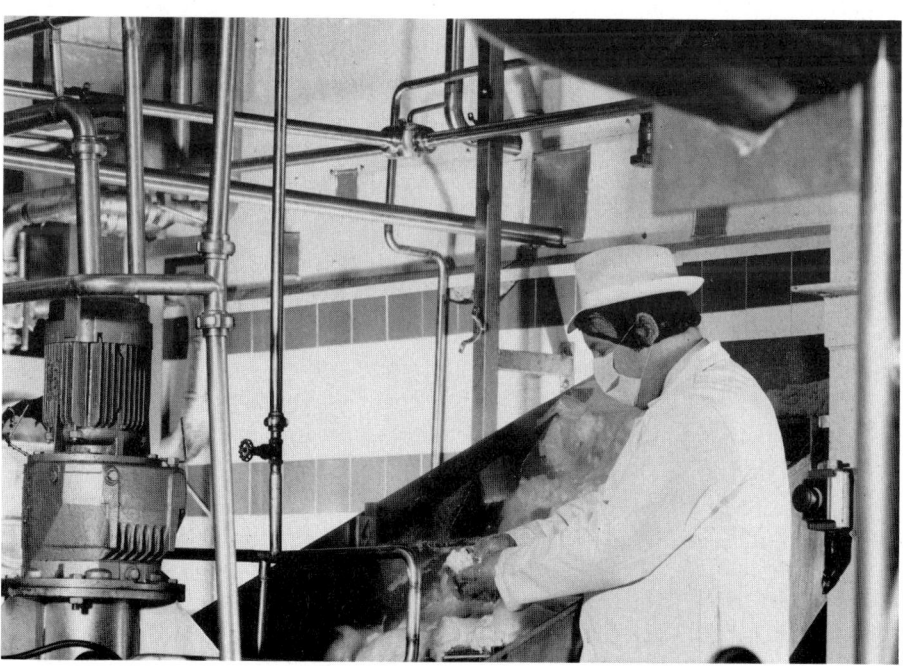

Hygienically testing casein curd: note hair, face and hand coverings

Mitchelstown Creameries

Organisation and personnel

This section lays down strict standards in respect of training and experience of all personnel involved in preparation of food and drugs. Rules relating to protective clothing, coverings for hair (including beards and moustaches) and hand and arm coverings as necessary are specified. Regular medical examinations by nurses and supervisors are necessary to prevent cuts, boils and infectious illnesses from contaminating the product. The regulations for controlled environment areas are especially strict.

Less stringent regulations apply to food preparation areas. Washing and sanitising of hands is particularly important, as is the constant cleansing of the areas and the utensils used in the preparation of food products. Canteen and other eating areas should be kept well away from working areas. Smoking is usually completely forbidden in the factory.

The GMP regulations insist that constant training and supervision are necessary to maintain the standards required.

Buildings and facilities

Production facilities should be designed to prevent contamination. The materials used in production are at risk from other possible contaminants, and are also themselves a source of potential contamination. So raw materials should be kept rigorously apart from finished products, and each stage in the process should be as self-contained as possible. The plant should be surrounded by properly drained grass and foliage, as a first-line protection against contamination. Access to the building by persons, animals (pets or vermin), insects, air, water, light and objects (including raw materials) should be strictly controlled. Double doors, air curtains and insectocutors should be fitted. Production and storage areas should have smooth, non-porous and easily cleanable walls, floors and ceilings. Temperature and humidity should be controllable, as should the air supply. Systems for cleaning and sterilising rooms and fixed equipment should be in place.

Extremely critical processing steps, such as the filling and sealing of sterile products, are performed in specially sterile rooms, called controlled environment rooms. In these areas the quality of the environment is precisely controlled. Standards for the proportion of dust particles, for floating microbes, for humidity and temperature are constantly checked and monitored. Procedures for entering and leaving such a room have to be strictly adhered to. Right of entry is restricted. Entry into such a room should only be through a gowning area, where clean (or new disposable) gowns and headgear and washing facilities are available. Shoes and personal jewellery should be left in the gowning room. The full suite of rooms in a controlled environment area consists of:
— gowning room
— air-lock

— sterile corridor
— holding area for sterile equipment, tools, product containers and closures
— filling and sealing rooms (which may be separate).

A critical aspect of the design of such suites is the control of flows of air and water. The suites are designed to change the air between twenty and thirty times an hour (as opposed to once an hour that would be normal in a warehouse, and perhaps fifteen times an hour in an office), and to help the control of temperature and humidity. The pressure in the air-lock is greater than outside, to prevent incursion. Water flows must also be tightly controlled in sterile environments. Floor drains are not encouraged; spillages should be wiped by wet/dry vacuum cleaner.

Exhibit: Building policy, Ballyfree Farms, Co. Wicklow
Ballyfree Farms specialise in the production of high quality cooked meat products, using turkey as the prime raw material. The various manufacturing processes are divided into separate and isolated units. The cooked meat, cooking and raw meat areas are specifically designated and three separate isolated units with interlocking electronic control systems prevent movement from one section to another. Movement of personnel from one area to the next is completely restricted to certain key personnel, who must adhere to specified procedures on protective clothing changing and sanitation. All personnel in the raw meat area are supplied with blue protective clothing. Orange protective clothing is used by cooking personnel, and white is the designated colour in the cooked meat plant. Each section of the plant has its own management and manufacturing teams.
Source: Ballyfree Farms

Equipment

In the drug industry, equipment should be qualified to work well within the degree of exactitude and the size of batch required. Its routine performance should be monitored and its efficiency checked periodically on a requalification schedule. The placing of equipment should maximise the efficient and secure flow of materials. Conveyors and transfer equipment should be designed to provide gentle handling of the containers, thus preventing hairline cracks which might let contaminated air into the product.

All equipment surfaces should be made of materials that are not reactive, absorptive or additive. Glass, stainless steel and Teflon are frequently recommended. Lubricants and bearing grease should be isolated from any possibility of their contaminating the product.

Cleaning and maintenance routines should be recorded in logs, and every aspect of the cleansing routine (date, times, operative, solvents and cleansing fluids used, test results) should be included in the log. Often it is desirable to remove equipment to a room specifically designated for cleaning. Special care should be taken with equipment surfaces. If they are scratched in cleaning, they may be impossible

Quality in Practice

to sterilise thereafter. Steam or high-pressure water cleansing can be used for tanks or pipes. Non-fibre-releasing filters should be employed in all sterile and non-sterile environments. The integrity of the filter should be checked frequently.

All equipment used for weighing or measuring components should be formally validated at least every six months. Scales and balances may also be checked daily by testing the reading for the smallest and largest weight within the range. Alarms, controls, pressure and temperature gauges should be frequently requalified. Laminar flow equipment, which controls airflow in specially protected environments, should be constantly monitored in operation. The rooms in which special airflow is required should be designed with non-shedding floor, wall and ceiling surfaces, and wall panels should be smooth and free from particle traps (such as screw heads, window frames, etc.).

Once again the requirements for the food industry are less stringent. The main requirement is a constant obsession with hygiene. (The details of food hygiene are fully covered in *Hygiene in Practice* by John A. Murphy.) All surfaces should be made of materials that can be maintained to prevent microbial infection. Stainless steel or non-shedding environments are not essential. It is critical that the devices recording the temperature of meat and other cooked products be calibrated regularly. If the meat doesn't reach the required temperature, the micro-organisms it contains will not be killed, and will multiply rapidly during subsequent processing.

Control of components and containers

Since there is no point in creating an elaborately sterile environment for the manufacture of the product and then storing and selling it in a substandard container, GMP demands close attention to storage containers. In general the company should develop written procedures for the receipt, identification, storage, handling, sampling, testing and approval or rejection of containers and closures. The need for emphasis on the closures was underlined by a case where deaths from septicaemia were traced back to micro-organisms lodged between the liner and the aluminium caps of a liquid drug product. Containers themselves should not be reactive, absorptive or additive in any way that could corrupt the drug.

Production and process controls

The instructions for the processing of a batch of drugs must be as specific as possible. Each significant step must be in writing: this includes the exact equipment and tools required, the components, the order of mixing, the process conditions (e.g. temperature, pressure, processing time) and the measurement of volumes and dimensions.

For foods, all the processes involved in receiving, inspecting, transporting, segregating, preparing, processing and storing of food shall be in accordance with adequate sanitation principles. The major risk is contamination with and stimulation of harmful micro-organisms. Every step in the process must be designed to minimise this risk. In order to achieve this, such physical factors as heat, humidity, time, pH value, pressure, etc. may need to be carefully controlled. In particular, food items that are liable to support rapid growth of harmful micro-organisms should be carefully temperature controlled.

Packaging and label control

Information on labels, whether for food or drugs, is carefully controlled by regulatory agencies. A normal drug label and instruction set will contain the name of the product, the in-house product number, composition details, type of dosage form, quantity in the container, strength, directions for use, counter-indications and side-effects, manufacturer's name and address, lot number and expiry date. Labels should be very tightly controlled in their storage and use. Only approved staff should have access to the storage rooms, and spoiled labels should be destroyed. Checks should be initiated to ensure that the right labels have been used for the batch. In sophisticated operations, the label contains a bar code which can be automatically scanned. With this technology, the label for every container can be checked during packing.

Warehousing and distribution

A warehouse for finished drug products should be neat, clean and orderly. The floors should be dry and dust-free; the lighting should be adequate. Facilities for storing special drugs should be specially designed. This includes refrigerated areas, and specially secure areas for quarantined and released lots of controlled drugs. The stock should be rotated on a 'first-in, first-out' (FIFO) basis, and adequate controls should be maintained to ensure that this is done. A typical system is to pick by lot or batch number, so that the lowest lot numbers are picked first. As far as possible the control of conditions should be pursued right down the chain of distribution. Special recorders can be purchased that will record on a disk or paper tape the temperature of the product and its containers during its journey from warehouse to retailer. The brand manual for Baileys Irish Cream, for instance, provides detailed instructions to retailers as to how the product should be stored and handled before sale.

Exhibit: Shipping Baileys Original Irish Cream
Baileys market and distribute their product Baileys Original Irish Cream on a world-wide basis. Consequently their product is exposed to all climates and to very variable warehouse, transportation and distribution arrangements depending on the country concerned.

The product itself is a cream liqueur containing fresh cream protected by alcohol. Being a natural product, the Baileys organisation goes to substantial lengths to ensure that special quality of naturalness, in taste and texture, is not lost through bad practices in the distribution chain. In some respects this is like those measures one normally takes to ensure that fine wines maintain their quality. For example, one stores wine on its side, i.e. in contact with the cork to allow the wine to breathe. In much the same way Baileys ensure that their product is not exposed to temperature variations or cycling and they go to great lengths to achieve this.

Thus, their Quality Assurance programme entails the use of temperature recorders to monitor conditions en route to overseas agents, regular market visits by Quality Assurance personnel, the purchase, as a matter of course, by their Marketing Executives of Baileys samples from across the globe and the subsequent analysis of these samples at their manufacturing facility in Dublin. Special brochures, posters and videos are prepared and distributed to overseas warehousing personnel to allow Baileys to guarantee to their consumers that the product arrives with them as fresh as the day it left Ireland.
Source: R. & A. Bailey product handling leaflet

Laboratory controls

Laboratory controls should include the establishment of scientifically sound and appropriate specifications, standards, sampling plans and test procedures designed to ensure that components, products, containers, in-process materials and labels

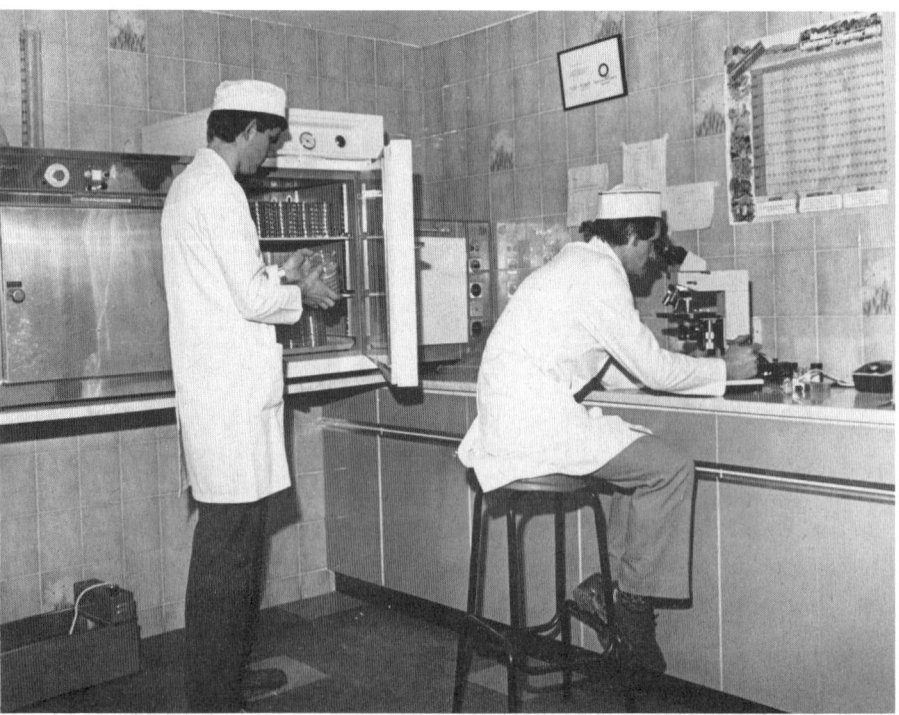

Monitoring bacteriological quality *Dairyland*

conform to the right standards. Controls initiated in the laboratory include the testing of all input materials, including retesting of any that risk deterioration in storage, testing of in-process materials instrument calibration, and testing of output. Finally there should be an established programme of stability testing to ensure that it conforms to standards throughout its shelf life. Sample batches of each production lot should be maintained and periodically tested for stability.

Records and reports

GMP regulations require that the firm maintain:
— records of the source and validation of raw materials sources
— records of all components, containers, closures and labelling
— manufacturing records of every batch of drug product, including evaluation records
— complaint files
— records of all retained samples
— records of salvaged lots
— records and documents associated with validation tests (including design and engineering of facilities, equipment, process and control procedures; and container/closure systems).

The documentation is usually divided into those elements relating to general procedural instructions (the quality manual and standard operating instructions) and those relating to the specific batch.

Other regulations

GMP regulations also contain detailed recommendations relating to returned and salvaged drug products, the quality of air and water in the plant, methods of sterilisation, and a comment on defect action levels for food.

Exhibit: Stimulating GMP awareness at Pfizer Chemicals, Ringaskiddy, Co. Cork
The Quality Control department has undertaken the responsibility of running an information programme to inculcate the philosophy that quality is a shared responsibility. The programme includes:
— Ensuring that the workforce at large is fully aware of the nature of the products we manufacture and their various usages. This has been achieved both by discussions in small groups (6-8 people) and by liberal use of our house magazine. (N.B. Do not assume that all employees know, really know, the nature of the products manufactured.)
— Informing employees of who the customer is, by name and business; what they use our products for, the kind of standards that they require of their suppliers and therefore the standards (especially of Good Manufacturing Practice) we must aspire to. Based on this, employees have been informed of what the customer looks for on his vendor audit, and why; equipped with this information the fulfilment of our GMP requirements by plant operators, craftsmen and technicians becomes more rational.
— Describing to employees the nature of customer complaints, the frequency with which they occur,

the markets where they arise, potential causes of the complaints and, above all else, requesting their suggestions for remedial action.
— Invoking the help of people at all levels to further improve our GMP, an easier task when the same people see and appreciate the reason for such an approach.
Source: J. Riordan, Director of Quality, Pfizer Chemical Corporation, Co. Cork

Conclusion

The risks associated with food and drug products are greater than with any other type of product. It is right therefore that the regulations relating to their manufacture should be stricter than for any other class of product. The emphasis in both the drug and the food areas is first of all on detailed and precise specification of standard operating procedures (SOPs). The SOP's first concern is to ensure that the environment, and particularly the working surfaces, are free of microbial and other contaminants. This goes right through the system, from the equipment, air control and personal cleanliness to the raw materials and the containers. The second aspect of GMP is to ensure that the batch or lot records are kept in every detail. This includes the records of all the tests and cleansing operations that control the environment, as well as the tests and controls on the individual batch.

Chapter 7

Just-In-Time production systems

ONE of the great challenges in managing a company is to isolate the areas where maximum profit or loss is made. Occasionally, re-analysing accounting or management information in a different way can provide startling insights. One instance of this kind of lateral thinking is the concept of quality costs, which adds together information from various budgets to produce a new cost centre. The full quality cost includes the cost of the quality department, the cost of quality training, the cost of warranty fulfilment, the cost of reworking and other aspects from across the budget spectrum. Another cost centre that could reward a special investigation is the value of stocks and work-in-progress. This is often treated as an irreducible figure essential to the proper running of the company, yet it is often so large an amount of money that it prompts special investigation to find out whether production can be arranged to reduce it. The cost of stocks and work-in-progress by themselves form only part of the story. The holding of such stocks involves the company in paying for:

— rent and rates for storage space
— heating and lighting
— insurance
— special costs (such as refrigeration for vulnerable goods)
— depreciation costs
— security and pilfering costs
— interest on capital.

The biggest item on this list is likely to be interest on capital tied up. At present rates, stock and work-in-progress of £100,000 will cost about £15,000 a year to finance in interest rates alone. The other charges between them could double this sum.

So why do companies carry stocks at all, if it costs them so much? The main reason is to provide a continuous service to their customers. Wholesalers and retailers, faced with an urgent demand for a product from the public, do not want to be told that there are production delays. The same reasoning applies to the internal planning of production. To get the best value from expensive machinery, it needs to be run continuously. This means, so the argument goes, ensuring that there is a queue of material waiting to be worked on at each machine.

Surveys show that for most of the time jobs are on the shop floor, they are waiting to be worked on. One study of the book printing industry showed that of the nine months that a normal book spent in production from manuscript to bound copies, at least six were spent in queues, waiting to be typeset, waiting to be proof-read, waiting for corrections to be done, waiting to be printed and finally waiting to be bound.

In order to prevent possible delays and to maximise machine usage, prudent production managers set up buffer stocks and queues. As a result, a vicious circle sets in. The longer a job spends waiting on the shop floor, the longer the queue will be. A long queue means that each job spends longer in production so it has to be loaded into the system that much sooner to meet delivery dates. This circle reinforces the need to have high warehouse stocks, because no customer will wait for the time required to work through the system. It also creates logjams, and queue-jumping for urgent orders, which between them make production control virtually impossible. This production control problem is aggravated when orders require several processing operations in succession. Different orders will require different times on each process. The cumulative time required on all processes for any one job can be as little as one-twentieth of the actual time spent in the factory.

Apart from the production control problems it creates, this system has a major quality defect. The queues and sub-assembly stocks actually conceal occasions when production isn't done right first time. This is of course one reason why they are there. The queue system ensures that if anything goes wrong with one batch, there are always plenty of back-up materials to work on. Nothing must stop the line. Buffer stocks diminish the effects of late deliveries, bad parts from suppliers, poor maintenance and other production problems. These stocks also have the effect of protecting one department or section from the mistakes of another. If the line doesn't stop, there is no urgency to meet and solve the particular problem. In effect, the buffer stocks paper over the company's fundamental quality problems, at enormous cost.

In the days when stocks and money were cheap, this may have been a good policy; but when companies can be locking up 30 per cent of their capital in stocks and work-in-progress, a different set of priorities comes into play. The high stock policy in effect conceals underlying quality problems. It is as if someone recommended solving the company's quality problems by bribing the quality staff not to report them. The method treats the symptom, not the disease.

The system called *Just-In-Time* production gets rid of the vast bulk of unnecessary stocks, and so saves money. It also removes the comfortable buffer that prevented quality errors from being quickly exposed. The old and the new systems can be compared to the way our meals are organised. In previous centuries (and to this day in countries like Nigeria), virtually all the items in the meal were put on the table at once. Soups, pies, roasts, stews and puddings all together. In the early nineteenth century the so-called Russian system became fashionable, whereby the meal was divided into courses which were served one at a time. When the soup was finished, the entrée was served and so on through the meal —

Good plant layout makes quality easier to achieve   R. & A. Bailey

everything placed on the table just in time. The main advantage was that every dish was served hot, but it also made the tables much tidier, cleaner and able to be decorated.

In factories just-in-time production works by eliminating both buffer stocks and the large batch production method. The idea is to produce each item just in time for use and not before. That is, make a piece, use it, make a piece, use it, and so on. The concept goes right through from final assembly to the suppliers of raw materials and parts. It works on the 'pull' system. When one station has nearly finished its work, it signals to the next to give it more. If the next area of production is full, the first doesn't produce anything until area two signals

Quality in Practice

readiness. In practice the signals are made just before the area is empty, to allow for delivery lead time.

Exhibit: JIT at Data Products, north Dublin

A typical JIT factory, such as that operated by Data Products in north Dublin, operates a bin or kanban (the Japanese word for bin) system. Data Products employees assemble computer line printers, serial printers and cheque printers. Parts are delivered from all over the world. The production system is demand driven, based on sales forecasts for the next eighteen months. From sales back through to stocks and suppliers, each section calls on what it needs from the next down the line. At a sub-assembly station, the parts needed for the sub-assembly task are stored in bins. Each station has two sets of bins, one set for the parts in current use, and one as the backup. When the first bin for any part is exhausted, a card is sent to stock, to order replacement materials, and the second bin is brought into action. Each part is treated individually, and each has its own set of bins. As a result of adopting JIT techniques of production, Data Products has moved its stock turnover rate from 2.5 times a year to 10 times a year.

Reducing setup cost

Just-in-time production reduces stocks by reducing lot sizes. Instead of making thousands of an item at once, regardless of the immediate demand, exactly as many are made as are needed *now*. This involves taking a new view of the arithmetic of the *economic batch quantity* (EBQ). In calculating the optimum amount to produce (or to purchase, the arithmetic is the same), there is clearly a play-off between the cost of holding stock and the cost of reordering it. Thus if a stock of drugs has to be kept secure and refrigerated, the carrying costs will be very high; in this case it is likely that little and often would be the batch formula. On the other hand, if the setup cost is high relative to the ongoing cost, the best plan would be to produce as many as possible in one batch. For academic books, for instance, the setup or 'first-copy' cost can be as much as 75 per cent of the total. Since the cost of storing books is not high, publishers print as many as they think they can sell in one batch, undeterred by carrying costs. The mathematical formula for discovering the EBQ was devised in 1917 by an American engineer called Camp. The formula is

$$EBQ = \sqrt{\frac{2NR}{C}}$$

where
N = the number of items required per period
R = the reorder costs per batch
C = the carrying costs per item per period.

The crucial figures in this calculation are the reorder costs per batch and the carrying costs. The reorder costs consist of the costs per batch of:
— adjusting stock records

— resetting and starting equipment
— batch inspection
— stores costs
— any rejects from startup
— change-over costs for the operators (new instructions, etc.).

Carrying costs consist of the cost per item of:
— rent and rates of stores
— heating, etc.
— insurance
— special costs such as refrigeration
— depreciation
— interest on capital invested.

Since the objective of just-in-time production is to reduce stocks and improve production efficiency, the first target is to reduce the reorder costs. Clearly the more often a batch is produced, the less the change-over costs for the operators will be, and the rejects during startup are an obvious target for quality investigation. The critical figure is usually the cost of resetting and starting equipment. Very often this is dictated by the technology. If a printer wants to change from red ink to blue, the printing machine has to be completely washed down. This is a lengthy process, but one which obviously has to be done thoroughly if scraps of red are not to appear where there should be blue. However, even in this case detailed investigation could no doubt discover inks or cleansing solutions that would allow the job to be done more quickly.

The statistics of production

Just as a JIT system should not be attempted until the setup costs for production have been reduced to a minimum, it should also not be attempted until the existing production process has been thoroughly investigated. Because JIT removes those comfortable buffer stocks, it is important to know what the true needs of the production process are. The core principle of JIT is that the output from process D falls into the hands of operator E just as E's output goes to the salespeople. The stocks in the bins and the relative production speeds at different stations have to be calculated very carefully, taking all the different production conditions into account, to enable this to happen as it should. This requires extensive use of statistical control charts (see Chapter 23). These charts monitor the natural behaviour of the various processes. First of all they display the rhythms of production over time, and then they expose the sources of non-conforming output, or variations in production. It is a major target of JIT production to increase the simplicity of the process. Control charts work towards that end by pointing out clearly when and how things go wrong. The key to successful introduction of JIT production systems is a detailed knowledge of the underlying production stresses and strains that is obtainable only through statistical process control disciplines.

Suppliers and JIT

Just-in-time production systems run through from the customer back to the supplier of raw materials. They have two objectives:
— to reduce the cost of carrying stocks, and
— to increase the quality of the product by doing little and often.

The importance of little and often delivery from suppliers was vividly demonstrated in 1983, at the height of the home computer craze. The makers of the brand leader Commodore 64 had a contract with a Japanese company which was to supply disc drives. An entire month's supply of 170,000 disc drives was found to be non-conforming, which stopped the sale of the machine completely at a critical time. Had the delivery been specified in lots of a few thousand at a time, the problem would have been quickly (and much less expensively) solved.

Obviously not every supplier can meet the very stringent requirements of JIT supply. The first step in starting such a system is to identify suppliers who are willing to cooperate. This may require them to initiate changes in their own systems (particularly in deliveries), and probably to set up statistical quality control systems of their own. The next step is to set up an evaluation system for suppliers, based on their past performance. The ideal is to reach such a level of confidence in the supplier that they can *ship-to-stock*, without more than a cursory evaluation by the purchasing company. This, however, requires detailed records of the supplier's past history with regard to quality and delivery.

The purchasing company's policy on choosing suppliers will be completely changed by JIT production. Proximity becomes a major factor in the purchasing decision, in order to reduce the cost of the many small-load journeys. In general, the company will look for a few very reliable suppliers, who can over time become closely integrated into the company's production process. The JIT philosophy aims at a different relationship with suppliers. The supplier becomes part of the production flow, not an impediment which has to be guarded against by high stocks. Finally, the quality specifications and standards have to be very clear. The supplier will be expected to meet those standards exactly, rather than being asked to aim for unrealistic standards in the belief that they are bound to fall short. Because the system is looking for massive savings on the cost of carrying stocks, it can afford to pay a little more to be sure that delivery and quality requirements are met. The benefits gained from a more rapid turnover of stock will greatly outweigh even a 10 per cent increase in cost.

Conclusion

Most production systems have an inbuilt tendency to accumulate stocks and work-in-progress. This is a result of equipment breakdowns, absence of operators, shortages and other temporary interruptions of workflow. It seems to make sense

Just-In-Time production systems

to keep operators and machines working, despite the fact that the output will not be used immediately. Equally it seems desirable to get work, any work, started as soon as possible. On the other hand, this method of working inevitably leads to queues, and to the masking of quality problems by developing buffer stocks rather than solving the problems at source.

As more money is tied up in queues and stocks on the shop floor, large interest payments have to be financed. Progress chasing and delivery schedules become difficult to maintain. From the quality point of view, this method of production control has the effect of reducing the significance and therefore the attention given to production and supply problems. If a company knows that a supplier is potentially likely, however occasionally, to deliver unusable stocks, the purchasing people will tend to build a buffer stock to prevent that having any effect on production. This solution simply covers over the supplier's weakness at considerable, though perhaps hidden, financial cost. Clearly the better solution is to persuade the supplier to achieve better quality standards.

Large batch large stock production systems make it difficult for the quality plan to have its best effect. JIT systems, on the other hand:
— reduce large-scale non-conformities that are a constant risk in large batch production
— enable the producer to learn about errors and to rectify them while the trail is still hot
— reduce the need to keep long-run batch production information, since the proof of conformity comes very quickly
— increase the detailed involvement of engineering and quality staff with day to day problems, by demanding regular problem solving attendance on the shop floor
— encourage the creation of quality circles and other worker-oriented quality awareness programmes by creating here-and-now solvable problems.

The principle of JIT is not suitable to all forms of manufacture, particularly those which have an irreducibly high startup cost. However for many systems it gives enormous benefits in financial terms, in production control, and in quality. Financial benefits arise from the massive reduction in stocks and work-in-progress. Production control advantages arise from the fact that the stream of production is greatly simplified by eliminated intermediate stocks, queues, and the long lead-times they imply. In quality terms the system ensures that quality problems are discovered and eliminated quickly. The batch production system risks the creation of large non-conforming lots. The pull system in effect tests every item as soon as it is produced.

Chapter 8

Information quality standards

IMAGINE a world without glass, or without gravity, or without light. Imagine a company trying to operate without anyone communicating with anyone else. Everything would stop. Without the constant flow of orders, requests, instructions, payslips, programs, complaints, standard operating procedures and stock lists all the machinery in the world is powerless. Nothing happens without information being passed from one person to another. The quality of that information is as important to the vitality of the company as the quality of the air we breathe is to our bodies.

But just as most of us never give a thought to how oxygen works in the body, the quality of information is usually taken for granted. It is only since companies have been spending large sums of money on computer information networks that serious attention has been given to the subject. Computers have in fact stimulated an important change in the way we think about information. Programmers have for years been trying to write programs which will enable computers to understand 'natural' language, such as you are now reading. As work on the problem went on, it became increasingly clear that messages derive a large part of their power from the recipient. Like an uneaten apple pie, a message without a recipient is not performing any function. The act of eating the pie makes sense of the whole cooking operation. Put another way, it's not what you say, but who you say it to that defines the message.

How information works

Information science began to be developed in the 1920s and 1930s in the Bell Telephone Company in the United States. Generalising from the model of a telephone call, the scientists described every act of information transfer as having six separately identifiable aspects. Some of these, such as the message, the source and the recipient, are obviously part of ordinary communicating. Others, such as the encoder, the channel and the decoder are more obvious in the context of radio or telephones than in everyday life. This is partly because we are so familiar with our language that we forget that it is in effect a code in which you can say some things but not others. Looking through a window, we forget about the pane of glass. Only when we listen to foreigners talking, or try to show the way to a tourist, do we notice the presence of the language code.

Exhibit: The elements of communication

Element	Telephone call
1. Source	Caller
2. Encoder	Caller's telephone set
3. Message	Conversation in electronic form
4. Channel	Telephone wires
5. Decoder	Receiver's set
6. Receiver	Person called

Computer programmers have a saying: 'Garbage in, garbage out.' This is simply a modern form of the old saying, 'There's no making a silk purse out of a sow's ear.' Since nothing happens in a company without the passing of information, clearly the way that information is passed is going to have a critical effect on the company's quality. A very high proportion of errors made in business can be assigned to faults in the information process. If the shop floor gets muddled instructions, it will produce sloppy work. If it gets clear instructions, there is at least a possibility of quality production. In planning the quality programme, therefore, anxious attention should be given to the quality of information. Each important document should be scrutinised with the six elements of information transfer in mind.

1. *Source*: The source of all information is the event as it occurs. Thus the source of a customer's complaint is a failed product, the source of a stock count is the stock in the warehouse, the source of an expenses form is the cash outlay. It is the place where the world of things is translated into information. The source of the information is the vital start of the whole process. The major difficulty with information sources is that, as a little-known aspect of Murphy's Law puts it: 'seven-eighths of everything is hidden'. It is too easy to take the appearance for the whole, for instance to assume that the stock count has been properly done, that the proper tests were carried out, to assume that the invoice has been properly calculated.

The quality questions to ask about the source of information are:
— is it what it appears to be?
— is it all that it appears to be?
— is it complete?
— is it in the right context?

2. *Encoder*: The problem with reality is that there is too much of it. In order to be able to handle the chaotic mass of 'that out there', the human brain has to define and summarise the key points. This is called encoding. Most of us have learnt to do this from childhood so it is automatic. Encoding has two aspects.

First the brain must interpret what it has seen. Thus if a product fails, the complaining consumer might describe this as 'stuck', or 'stopping', or 'not going'. Each of these might imply a subtly different fault.

The second step is to summarise the seen information into coded form. Codes, or languages, are more or less effective in different contexts. Take the codes relating to colour. Most of us have a very limited colour vocabulary of perhaps ten or twenty basic terms. Printers, on the other hand, distinguish over 600 shades, referring to them by a code number. An even more elaborate standard definition distinguishes 267 basic colours and over 7,500 shades. Thus a colour that a layperson would describe vaguely as 'darkish purple' might be referred to by a printer as Pantone 260, or by a paint manufacturer as African violet. The reality is the same, the code is different. Codes allow the informant to be more precise about what has been experienced. Each environment develops its own code or jargon. Thus the Arabic language is rich in words to describe various ages and types of camel, and the Eskimos can distinguish twenty or thirty different types of snow. In the business context, communication requires that, for instance, quality standards be expressed both in the full-code technical terms and also, if necessary, in simplified terms for the shop floor and senior management.

Quality questions to ask about the encoding are:
— did I understand what I saw?
— have I used the right code to describe it?

3. *Message*: The message is the parcel of information that goes from sender to receiver. It has two aspects: the substance and the context. The context is provided by the recipient of the information, the substance by the sender. In business terms, messages serve one of two functions:
— to reassure us that everything is going well (e.g. sales, temperature, productivity, quality conformance levels, etc. up to target), in which case production and other activities run on as before, or
— to alert us to a change in a key variable.

The latter is the important one. Indeed the former could be dispensed with, were it not that humans like to be reassured. Just as motorists like regular signposts to confirm that they are still on the right road, managers spend much time and money reassuring themselves that all is well.

The message is an encoded summary of what the sender knows about the subject. The quality questions specifically relating to it are:
— does it say what I want to say?
— does it say everything and only everything?
— how does it relate to the context?

4. *Channel*: In most cases the specific channel of communication is part of a general system of communication. Thus most companies have internal telephones, internal memo systems, special forms and standard report formats. There are

Information quality standards

Clear labelling of areas in the factory communicates the quality message
Thermo King, Digital

advantages and disadvantages to this kind of standardisation. The great advantages are that the standard form is well known and understood, so the recipient wastes no time understanding how the channel works before grappling with the message. The importance of this point is obvious to anyone who has tried to send or receive messages on an unfamiliar computer network. Familiarity enables the eye to run quickly to the crucial bits of information and digest them. Standard forms are also very useful timesavers from the sender's point of view. They act as a checklist of the types of information required, and no time is spent on designing a new format every time a routine piece of information is sent.

Channels lose efficiency through what is technically called *noise*, which is the corruption of the message by extraneous matter. The term comes from the way the crackling on a radio interferes with the reception. A familiar example of channel noise is when a verbal message is carried from one person to another and subtly rephrased by the carrier. The new words can often convey meaning very different to that intended. Other examples of noise are when an appliance instruction contains too much technical information to be understood by the ordinary user, or a road sign gives the distance of so many places that it is impossible to find one's own destination quickly. Because of noise, most codes are designed to carry *redundant matter*, so it doesn't matter if some of the message gets lost. English spelling, for instance, is full of redundant letters. Mst ppl hv lttl dffclty n rdng ths sntnc. Btrmvngthspcsmksthrdr. Removing vowels and spaces reduces the message to less than half its length, but makes it necessary not to lose a single element, or the message would be impossible to decode.

The quality questions to ask about a channel are:
— does it help or hinder the transmission of the message?
— which is the best channel for this message?
— can the noise in the channel be reduced?

Exhibit: Information system at Merck, Sharpe and Dohme, Co. Tipperary
At the Merck, Sharpe and Dohme bulk drug manufacturing facility at Ballydine in Co. Tipperary, a computerised Plant Information Management System allows the plant operating personnel to rapidly evaluate the effects of changes in process parameters. The system comprises a DEC VAX 11/750 minicomputer with twenty-four terminals and ten printers.

There is a danger with all information systems that they can suffer from 'data pollution', i.e. too much data and not enough information. Therefore it is necessary to provide the software to interrogate the data and produce information. The system collects and stores electronically the following information relating to the production of bulk pharmaceuticals:
— critical temperatures, pressures, levels, etc. recorded during specific production operations
— quality results from tests carried out either during the production operation or on product samples
— the temperatures, pressure levels, etc. from all the vessels used to manufacture the product. All the data values are recorded each minute and the system stores sixty days of data on line for instant retrieval. Also there are two years of data stored off line on magnetic tape
— the lot numbers and quantities of all materials used in the production process.

The data stored electronically can be reviewed either in tabular, graphic or statistical form. The tabular reports are structured so that the data can be reviewed in either summary form for management review or in detail form for problem investigation. Graphically the data can be reviewed as either linear plots of quality or process parameters versus batch number or as scatter plots of quality and/or process parameters plotted against one another. This allows trends or relationships between parameters to be identified.

5. *Decoder*: The decoding function is obviously necessary in some cases, such as a letter from a foreign customer, a computer program, a balance sheet, or a sophisticated control chart. In these cases most people will need to be told what is going on. They need to be told the code. Other forms of communication are less obviously coded, because the code is so familiar to us that it has become transparent. It is only when we try to speak a foreign language, or even listen to a group from another industry that the importance of the code reasserts itself. The same applies to technical codes such as statistics, computer programming languages or the conventions of accountancy; for some people they become so easy to manipulate that they present no barrier to communication. For others they are a major obstacle.

The quality questions to ask about the decoding function are:
— has the decoding process got the information out of the message?
— is the signal received now the same as the source intended?

6. *Receiver*: The purpose of a message is to change things, that is to stimulate action, or to confirm results. But any message makes sense only in a context. Suppose for instance a printer's rep rings the sales office: 'O'Brien's want 200 of the 005s by next month, and Cashman wants another 15 of the 101s.' The sales clerk knows that in the publishing industry it is customary to drop the thousands when talking of print quantities, so 200 means 200,000. On the other hand the 101s are leather-bound ledgers, of which no one will want 15,000. The context, understood by the sender and the recipient, makes sense of the message. Suppose again that O'Brien's usually ordered 500 of the 005s every month, or conversely had always ordered that product from a competitor — the context radically changes the impact of the message.

A major source of misunderstanding in communication occurs when the context envisaged by the sender is different to that seen by the recipient. The famous Charge of the Light Brigade was caused by a message sent in one context and received in quite another. Lord Raglan stood with his staff on the top of a hill, from which he could see two groups of guns, one easy to attack and one, the Russians, virtually impossible. He sent a message to Lord Cardigan 'to attack the guns'. Lord Cardigan was in a valley, and could see only the Russian guns. To attack them meant a frontal attack on a well entrenched position. He had his orders, however, and he and his six hundred cavalrymen charged the guns. They were massacred. As Tennyson put it, 'someone had blundered'.

Clearly it is the recipient's context that is critical, since it is he or she who is going to react. It is the responsibility of the sender to ensure that the contexts are seen as the same.

Quality questions to ask about the receiver are:
— what is the context in which this message will be received?
— how will the message be reacted to?

Conclusion

Companies run on information just as cars run on petrol. If the information handling is inadequate, the company will eventually grind to a halt. In order to investigate the quality of information, the six elements in the communication process should be used as quality characteristics. Every communication, whether it be a quality standard, a public notice, a batch record, an internal memo or a sales brochure, must be designed with the six elements in mind.

In most cases the three critical aspects will be the source, the message and the receiver (particularly the context in which the message is received). Obviously a message based on incomplete data received in a different context to that in which it was sent could have disastrous results. However, the more technical aspects of communication theory, the coding and decoding and the channel, should not be ignored. Is it better for an eye lotion to ask the user to 'irrigate the eye', which is the technical expression, or to 'wash the eye with cold water', which everyone will understand? The vulnerability of verbal messages to corruption in transfer (channel problems) is proverbial — it's said that one urgent plea for reinforcements before an advance during the First World War arrived at the Staff HQ as 'Send three and fourpence, we're going to a dance'!

The processes of production are controlled by information flows round the company. In the long run the quality of output is critically affected by the quality of that information. One is not possible without the other.

Chapter 9

The customers' contribution to quality

CUSTOMERS make their contribution to the quality process by complaining. They complain about products because they are not satisfied. They are not satisfied because they were expecting something else. That expectation was more than likely created by the selling company's advertising and marketing. The product offered failed to match the expectations created in terms of service, delivery or fitness for use in some way. Price is only a problem when the customer feels that the quality delivered doesn't match the price paid. Price is known in advance, so cannot in itself be part of the problem of frustrated expectations.

Exhibit: Attitude to the customers
A customer is not dependent on us — we are dependent on her. A customer is not an interruption to our work. She is the purpose of it. We are not doing her a favour by serving her. She is doing us a favour by giving us the opportunity to do so. A customer is not someone to argue with or match wits with. Nobody ever won an argument with a customer. A customer is a person who brings us her wants. It is our job to handle them profitably to her and to ourselves.
Source: Extract from Marks and Spencer Trading Policy

Irish consumers are traditionally reluctant to complain. This is not unique; reports from the US suggest that only 2 per cent of customers will bother to complain about low priced goods, and perhaps one in twenty will complain about goods costing a few pounds. The main reason for not complaining is that it isn't worth the trouble: the customer usually decides that the most effective form of complaint is not to repurchase the product. Other reasons include the dislike of hassle, perhaps encouraged by the negative response to previous complaints, and the easy-going toleration of failure as a fact of life. Customers are less likely to complain about one-off failures in products they use frequently than about a failure the first time they use a product.

Research conducted by the American Society for Quality Control suggested that for every complaint received, there are at least six seriously dissatisfied customers, and between twenty and fifty less seriously dissatisfied customers. The more expensive the product, the more likely it is that dissatisfied customers will complain to the company. But just because dissatisfied customers don't tell the company, it doesn't mean they keep the bad news to themselves. Dissatisfied customers discussed the product with at least twice as many people as satisfied

Quality in Practice

Discovering the customer's point of view *Mitchelstown Creameries*

customers. Even of those who actually complained, those who were happy with the company's response told five people, while those who were not happy told ten.

A survey in Ireland in 1980 suggested that causes for complaints, whether voiced to the company or not, are frequent. Of those who had bought shoes or electrical apparatus in the previous year, 9.5 per cent had complaints; food/drink/tobacco, clothes and TV sets generated between 5 and 6.5 per cent complaints; new cars and hi-fi equipment 3.5 per cent.

The rest of this book describes systematic ways of ensuring that the quality delivered to the customer maximises fitness for use. Occasionally however the best laid plans go adrift, and customers complain. This should be approached positively. By actively encouraging complaints, a good complaints handling system can generate various different benefits. These might be:

1. *Extra profits*: Only 1 per cent of people complain about products costing a few pennies, and even for those costing £200 only 50 per cent complain, despite dissatisfaction. Everyone, however, tells their friends of the problem. The one complaint that actually gets through to the company thus represents the tip of the iceberg; in a low priced item, as many as 200 people (99 non-complainers

telling 2 friends each) may receive a bad image of the product and the company. Each of these is likely to be a potential user of the product, since only fellow users would find the matter of interest. If the company encourages complaints and makes it easy to find redress, the story can become a good news one. Another US survey, this time sponsored by the White House Office of Consumer Affairs (the TARP Report), found that 70 per cent of those customers whose complaint was satisfactorily resolved purchased the product again, as opposed to only 46 per cent of those whose complaint was unsatisfactorily resolved and 37 per cent who did not complain at all. Even for items costing more than $100, over half of those whose complaint was resolved purchased again; only 19 per cent of those dissatisfied with the complaint handling did so, and only 10 per cent of those who did not complain. The non-complainers simply vote with their feet; they don't give the company a second chance.

2. *New product ideas*: Complaints reflect ways in which customers are dissatisfied with the product; it has failed to come up to expectation. This is an obvious pointer for the product development team.

3. *Product performance information*: Complaints can also reveal the way that the customers are actually using the product. If this isn't as intended, then the company has failed to instruct the users properly.

4. *Improved relations with retailers*: Most retailers are not equipped to deal with complaints, and dislike being blamed for selling unsatisfactory goods. They will be favourably biased towards companies that take the trouble to make it easy for customers to complain directly to the manufacturer.

5. *Improved company image*: A good complaints handling system can generate loyalty from the complainants to the company, which will spill over into purchase of the company's other products.

6. *Better educated customers*: Undoubtedly many complaints arise from customers misusing the product, or simply failing to read the instructions. If the complaint service is able to point this out tactfully, most customers will respond positively both by reading instructions a bit more carefully in future and perhaps by increased brand loyalty (at least they have found out that the product works!).

The research that has been done on the effects of customer complaints has been related to large companies making consumer goods. The same message certainly applies to small companies making industrial goods. In every case the challenge to the company is to use the complaint to turn a dissatisfied customer into a satisfied one, and to use the 'free market research' to best advantage. Complaints must be dealt with promptly and effectively, so that the customer's confidence in the company is restored. A complete complaints handling system should contain the following elements.

1. *Generating complaints*: It may seem strange that companies should encourage complaints. On the other hand if complaints are treated as free information and market research, it is intelligent for companies to encourage them. The name and address of the complaints manager should be clearly visible on the instructions and wrapping (and not buried on page 73 of the manual). Many companies in the US give reverse charge numbers for complaints. Dissatisfied customers often fail to complain because they don't know where, or how, to do so. Another reason is lack of faith that the company wants to receive complaints. If this information is clearly visible, potential complainants know what to do; satisfied customers are reassured at the same time.

2. *Input system*: Companies receive a wide range of general information and communications: invoices, orders, payments, requests for information, answers to letters and complaints. The first step in the complaints handling system is to ensure that complaints, whether they come by post or by telephone, are channelled as quickly as possible into the system. The second is to log them in a daybook, so that a record is made of their receipt. A major cause of dissatisfaction is the failure of companies to respond to complaints. The logging enables the response rate, which should be a major objective of the complaints handling system, to be monitored. Finally, the complaints should be classified. One object of classification is to ensure that those which might be symptoms of a serious product liability problem are 'fast-tracked'. The classification system should serve as the basis both of action and of the report system to senior management.

Exhibit: Bailey's customer/distributor complaints handling system
17.2.   *Objectives*
17.2.1. All complaints are rapidly and effectively handled in accordance with company policy.
17.2.2. The customer's difficulty is alleviated promptly.
17.2.3. Same problem does not occur again because the cause has been identified and corrected.
17.2.4. The product rejected by the customer is returned for analysis to our Technical Department.
17.2.5. Customer's confidence is restored in our product.
17.2.6. The relevant information is recorded and reported for management.

17.3.   *Policy* Our company policy is to replace any product rejected by the customer.

17.4.   *Responsibility* The responsibility for running the Customer Complaints Handling System is shared by Customer Services Department and Technical Department, as agreed with Senior Management.

Procedure
17.4.1. *Registration* All complaints are channelled to Customer Services Department and recorded in a log, and assigned serial numbers to assist in tracing.

17.4.2. *Data Collection* On replying to complaints we request that the product is returned to us for analysis. We also request information as to where the product was purchased.

17.4.3. *Complaint Grading* – *attached* Complaints are graded into one of three categories, critical, major and minor, using previously established guidelines.

17.4.4. *Customer Contact* All contact with customer is handled by the Customer Services Department.

17.4.5. *Reports* Reports will be issued by the Technical Department on a monthly basis for review at Management Meetings.

Source: R. & A. Bailey Quality Manual

---

3. *Response system*: The classified complaints must now be investigated and responded to. Normally complaints will fall into well worn paths, but occasionally special investigation will have to be undertaken. In this case it is often necessary to ask the customer for more information (typically the company may need to see the packet or the wrapping to check the manufacturing batch records). Even if it isn't necessary to recontact the complainant, it is usually advisable to do so in order to make time to solve the problem. The response can be in the form of predrafted letters into which the special details of name, address, etc. can be slotted.

An important part of the response may be the replacement of the complained-of product, and certainly a refund of postage. The system should be set to ensure that the standard letter saying, 'I enclose a replacement etc. and a refund' does in fact do so.

4. *Internal Corrective Action Request*: The value of the complaints system is that it provides objective information on how the customer perceives the quality of the product. To get the benefit from this input, the company must have a good Corrective Action Request system. In many cases the complaint will refer to a known problem; in others it may be the first alarm call about a serious change in the delivered quality of the product. Perhaps the transport manager has changed hauliers, and the new people are less careful with the product than the old; perhaps a new supplier has been found for a key component, and the proportion of non-conformance, though within tolerance, is higher. In the latter case, the acceptable quality level may need to be tightened. The Corrective Action Request should come with the authority of the quality manager, and should be a priority message. A copy should be kept by the quality manager, and the result followed up regularly.

5. *System management*: The complaints system should be monitored like any other. Typical points of measurement are
— analysis of complaints by product (type, area, seriousness)
— analysis of complaints by raw material suppliers
— speed of response to complaints by letter/telephone
— warranty, and other complaint handling costs.

Another aspect of managing the system is the motivation of the staff. Complaints handling can be a stressful job, especially since the complaining customers often express themselves very aggressively. This is in part to overcome their own nervousness about making the complaint at all. The following guidelines can be used to provide a positive attitude to complaints:
— with aggressive or rude complainants, listen and wait: provide sympathy, help and an active response
— never treat a complaint as a personal insult to the company or any of the workers
— do not treat complaints as something that stops you getting on with your work: the customer is your work
— always treat customer statements as honest expressions of how they see things; people should not be treated as potential con artists
— always acknowledge the complaint and treat it seriously from the beginning
— advise the customer how the complaint will be dealt with, and how quickly
— follow up and carry out these commitments
— if the decision is not what the customer is seeking, give clear logical reasons for it.

Conclusion

However good the quality management, it is likely that there will be complaints from customers. There will also be dissatisfied customers who do not complain. It is to the company's advantage to encourage dissatisfied customers to complain, thus giving itself a chance of turning a dissatisfied customer into a happy one. Also, by using complaints as an independent source of information about the product, the complaints system operates as free market research. To do all this the quality plan must include systems for making it easy for the customer to complain, and for ensuring that once a complaint has been made, satisfaction will be rapidly forthcoming. The second objective can be achieved by recording and analysing the complaints that come in, and by re-examining the design and production of the product in their light.

Chapter 10

Service without a snarl

THE service industries are a major part of every developed economy. In the US it is estimated that 60 to 70 per cent of the working population is in service, as opposed to agriculture, mining or manufacturing. Typical service operations include hairdressing, computer software, public transport, schools, banks, airlines and hotels. There are five factors that render them different from manufacturing industry.
 — *Intangible output*: You cannot transport the output of a service industry.
 — *Perishability*: The service must be produced on demand; it cannot be stored in a warehouse.

Well-planned warehousing contributes to quality: each batch of ingots is self-palletised to reduce tare: it is also shrink wrapped *P. Carney*

— *Complex delivery system*: Services are usually delivered in very complex environments (e.g. airports, hotels, banks) that require elaborate and time sensitive management.
— *Customer presence*: The delivery system also includes the customer, which adds a major element of unpredictability.
— *Customer standards*: The performance of the service is judged by the customer's subjective evaluations, which are often difficult to predict.

Quality characteristics

In the manufacturing environment, quality is often a matter of exact measurement. Specified tolerance levels are achieved, or not, and can be precisely measured. Consumer goods are either fit for use or not. Medical equipment is either sterile or not. The output of the service industries is not so easily quantified. This is because people are involved, and the things they require from a service are not measurable. How, for instance, can you assess the friendliness of a bank? Or the stylishness of a restaurant? Or the attractiveness of a hairdo? Because of these difficulties, service industries have tended to resist the methods of quality control that work well in engineering environments. It is even argued that if the service can't be standardised, the methods of quality management cannot be applied to it.

The problem boils down to a search for quality characteristics that can operate as *control subjects*. A control subject is an indicator that reflects changes in the service being provided to the customer. Like body temperature, or the Consumer Price Index, it is a single figure that reveals general changes for good or ill. It must be reasonably accessible, and measurable. (See Chapter 11 for more about control subjects.) Some writers have identified speed and timeliness of delivery as being *the* quality characteristic special to the service industry. Every service industry has to worry about this, because a service by definition is delivered 'on the spot', and cannot be stockpiled (you cannot keep a stock of haircuts in case there's a rush). Time as a quality characteristic has three aspects.

1. *Access time*: The length of time it takes to gain the service company's attention. Standards relating to this might be expressed as 90 per cent of calls should be answered within 15 seconds of first ringing, or no caller should be transferred to more than one number without being asked if he or she would prefer to be rung back.

2. *Queuing time*: Once the client has gained the company's attention, he or she has very often to wait to be serviced. The customer will be concerned with the length of the queue, the fairness of the queuing procedure and the conditions of queuing. The situation to avoid is represented by draughty old-fashioned post offices, where separate and often long queues are formed at different windows. A corollary of Murphy's Law states that whichever queue you join, that is the

one that will contain the three customers who are respectively: (1) claiming old age pensions for three cousins, at least one of whom died recently, (2) taking out small sums of money from five separate post office accounts, and (3) sending a parcel of eggs to friends in north China. Most banks now have one queue served by several tellers, which solves the fairness problem.

3. *Action time*: Having finally reached the head of the queue, how long does it take to complete the business?
Each of these aspects is important. They have a cumulative effect on the impression given and the quality of the service; the longer the customers have to wait for access and queuing, the more exasperated and difficult to please they will be when finally attended to.

Exhibit: Aer Lingus' Quality Quest programme
The passenger view
If Quality Quest is about anything, it is about giving a better service to the customer. Therefore customer satisfaction is the essential measurement of our performance and progress. This can be measured in two ways: firstly passenger complaints, and secondly passenger compliments. Both of these measures show favourable trends over the past year. Passenger complaints have declined by 20 per cent, ... compliments have increased by 50 per cent. Last year we introduced over fifty separate improvements to our service ... [however] we need to recognise that punctuality is the single most important measure of our quality of service. Last summer, 20 per cent of flights were delayed for less than fifteen minutes.
Now for the good news
Fact: On-time punctuality at Dublin Airport improved by 18 per cent in the year to March 1985.
Fact: We are currently ranking within the top five of fifteen airlines in Europe.
Source: Aer Lingus Quality Quest Action Group, June 1985

The key to the value of the service lies in the customer's judgement. This raises the further question of exactly when the service is seen as finally delivered. Some services such as dentistry are judged over a long time after delivery (the longer the better), and some as they are delivered (such as transport). In some services matters unconnected with the service itself may be important: builders and repairmen are rated by careful householders as highly on the mess, or lack of it, they leave as on the work itself. Another approach envisages the customers separating in their judgements the *hard* elements in the service, the technology, from the *soft* elements, the manner in which the service is performed. The customer may judge that the technical service was fine (the plane arrived safely and on time), but the manner was off-putting (no boiled sweets, no free coffee, etc.). Banks, financial houses and airlines often find that because customers can scarcely distinguish the technical services performed, they must advertise their superior 'friendliness'.

In this way, behaviour itself becomes a quality characteristic. Doctors have long known that a good bedside manner is more highly valued by patients than any

Quality in Practice

	PROFICIENCY		TIMING	
	Degree of seriousness		Degree of seriousness	
	Most serious	Less serious	Most serious	Less serious
Hospital medication	Wrong drug given Wrong amount given	Dose omitted	Dose given at wrong time	
Post	Letter or parcel lost	Letter or parcel defaced or damaged	Undue delay in delivery	Delivered at wrong time of day
Restaurant table service	Table dirty Spillage on guest	Wrong dish served Unwelcoming attitude by staff	Undue delay in starting meal	Undue delay between courses
Retailing: food	Out of stock of basic items	Bad attitude of staff	Delays at checkouts Trollies not available	Poor service at depts
Retailing: drapery, hardware, etc.	Poor product knowledge	Using incorrect wrapping materials	Not serving customers in turn Goods not in store on date promised	Delays while goods are being located in stockroom
Road transport	Goods lost	Damaged in transit	Goods not delivered on time	Goods delivered during meal times
Coach transport	Poor mechanical condition of coach	Dirty exterior of coach	Late picking up of passengers at start of journey	Late arrival at destination

Examples of measures of service quality IS 305: 1985

Service without a snarl

	PROFICIENCY		TIMING	
	Degree of seriousness		Degree of seriousness	
	Most serious	Less serious	Most serious	Less serious
Freight forwarding	Onforwarding consignment to wrong destination	Removing exporter signs from containers	Failing to phone clients on time as promised	Undue delay in issuing invoices
Shipping agency	Bills of lading not complying with shipper's instructions	Not sending bills of lading in ship's bag as requested	Not issuing bills of lading in time	Failure to advise exporters of delay of vessel
Typing and duplicating	Loss of manuscript or other input e.g. shorthand notes, recording tape, etc.	Typing errors Poor layout Copying errors	Work not ready on time	
Banking: current accounts	Non-payment of standing orders Wrong amount paid	Unhelpful attitude by counter staff	Standing order paid at wrong time	Undue delay in getting counter service
School teaching: primary school	Failure to maintain control of class	Pupil's loss of interest Poor conveyance of knowledge	Curriculum not completed in scheduled time	
Dry cleaning	Customer's garment lost Garment damaged	Removable stains not removed Poor attitude by counter staff	Work not ready on time	
Emergency ambulance	Patient suffers further injury by movement		Late arrival at place of accident	

number of degrees; the human characteristics of the service are a vital element in its acceptability. Another important quality characteristic to service industries is image. In many cases the image of the service is as important as the service (e.g. fashion products such as clothes, night clubs, restaurants). In this situation, the factors which maintain the image must be included in the quality control system.

Measuring service

The first stage in introducing quality assurance into a service industry is to choose the relevant quality characteristics. The second stage is to identify control subjects that will represent significant changes in the value of the service. The next stage is to set the standard. In some cases, such as in food hygiene, there will be legally enforced standards to be met. In most cases, however, the company must set its own standards. An airline might divide its standards into sets relating to operational characteristics and passenger service characteristics. Operational characteristics could include the proportion of flights that actually took off as a proportion of those scheduled, the proportion that were on time, the proportion of flights that had operational problems inflight, etc. The passenger service standards might include the number and type of passenger complaints, the number of baggage problems per 1,000 pieces handled, etc.

In other contexts, the speed of service might be important. An Post sets itself a target of delivering 90 per cent of letters by the next working day. Research by the Consumers Association of Ireland, based on a sample posting of over 1,000 letters, suggests that this target is met for letters within Dublin, but for letters between Dublin and the rest of the country the next day delivery rate can drop as low as 75 per cent. In a normal queuing situation, the speed of service as a quality characteristic could be monitored by giving every customer a time-stamped and numbered ticket on entry. Each number is called for service in turn, and stamped with the time of service by the service clerk. The tickets can then be monitored for such factors as average waiting time and queuing cycles (rushes and slack periods).

The choice of quality characteristic can be complicated by the fact that few services are the result of a single activity. Most are combinations of several elements, quantifiable and non-quantifiable, but if one aspect of the service is less than perfect, it can often be compensated for by others. The final standard is the customers' judgement. Of course the customers will usually be unable to quantify their tastes (as in 'I don't know much about art, but I know what I like'). In this case the service manager has to set arbitrary standards, and then see if that level is acceptable to the customers. A pitfall to avoid is that of setting an easily measurable standard that is important in itself, but is only part of the picture. The temptation then would be to achieve the highest possible levels of that standard, at the expense of other desirable objectives. It would be very 'friendly' of a bank manager to give very easy overdrafts, thus fulfilling the objectives of the public relations policy at the expense of the profit and loss account. Unfortunately, not even the friendliest building society in the world would do that.

Once the standard has been set, conformance with that standard must be checked. The ordinary industrial techniques of measuring and testing are often useless here. There can be no tasting panel to judge the flavours of the restaurant meal — the chef's skill is the only criterion. Another complexity enters in the person of the customer. Every customer is different, so the service has to be, at least to some extent, tailor-made. The customer is also a relatively unknown variable. His or her actions and tastes can significantly affect the quality of the service, in ways that are quite out of the control of the supplier. The system for delivering the service has to be designed to take account of the variability of the customers' demands, and to be able to cope with them. The quality assurance system must plan to cope with the vagaries of the customers. One essential part of this is the contingency or fall-back plan. If fog hits the airport, or acts of God or people prevent the service being provided in the expected way, it is not enough for the service staff to wring their hands in despair and go home to bed. Alternative methods of providing at least an adequate substitute have to be planned for.

One way of establishing and maintaining quality is to ask for feedback from customers. Many hotels and other organisations now ask customers to complete a form expressing their satisfaction and dissatisfaction with various aspects of the service. Just as an effective complaints handling system can help a company develop its product and gain valuable information on how the customer perceives the product, so a feedback system can help monitor the level of service delivered. It is of course vital that, if a customer takes the trouble to fill out a suggestion/complaint form, he or she should be answered. A complaint ignored multiplies the ill effects of the original fault.

Many parts of the service industry simply cannot tolerate less than complete accuracy. The only acceptable level of customer safety on an airline is 100 per cent. It is not enough for a hospital to give the patient the right prescribed drugs *most* of the time. This demand is made more stringent by the fact that the product is delivered to customers as they stand there. There is no checking at the factory gate to confirm that quality standards have been met, and no bulk buying so that failure in one or two items in a batch can be rejected. In fact there is often nothing for the customer to reject. If a lawyer bungles a lawsuit, it is too late to complain.

In this case, as so often with service industries, the sales staff are the same as those providing the service. There is no division between the functions as is typical in manufacturing. This has implications for supervision and control. Obviously if the service is delivered from scattered points, it is impossible for inspection and testing to cover them all. What can often happen is that standards or performance become weak, because they are set by the individual workers.

The only approach in this context is to fix standards as required by management and to train and motivate the employees so that the desired standards become their own.

Conclusion

The special nature of service industries gives rise to particular problems in adopting quality planning. Most quality disciplines were developed to handle normal industrial circumstances, where products are designed, made and inspected before being despatched to the eventual user. In service industries the product is effectively created and consumed in the same action. The customer is a vital part of the operation. The final judge of the value of the product is the customer, and he or she may be unable to provide objective standards by which to judge the product's fitness for use.

Even if we can identify the critical aspects of the service, they often reside in such unmeasurables as stylishness, friendliness, sensitivity and human warmth. On the other hand staff can be trained, and systems designed, to make the best of relations between customer and staff.

An important part of quality planning for service industries is to set up good communication channels between the customer and the suppliers of the service. This enables the suppliers of the service to remain sensitive and responsive to the customers' needs.

PART II

Pride in the product

Chapter 11

Making the system work for quality

A business is many different things at once to the people who are affected by it. For the customer it is a source of goods, and for the shareholders a source of dividends. For the employees it is a place of work, a meal-ticket, a social centre, and a focus of identity, of ambition, of creativity. At one level it is a gathering of bricks, mortar and concrete, housing masses of metal and paper animated by power and people. At another it is a highly complex arrangement for efficiently gathering in raw materials and making and selling products.

One way to understand these conflicting levels, and to be able to compare one company with another, is to talk about them as *systems*. This is a way of describing operations according to how they are controlled. For instance, sometime in October every year Bridget starts the central heating. She decides on the temperature to be maintained, and fixes the thermostat accordingly. As long as the boiler is lit, the thermostat will test the room temperature, and if it gets too warm, will instruct the switching mechanism to turn the boiler off for a time. When the temperature drops, the sensors instruct the switch to reactivate the boiler. A clock device adds a second condition to the controls, by turning the switch off during the day while she and her husband are out at work. No matter how low the temperature, the boiler won't come on until the due time.

This familiar mechanism illustrates the operation of a simple control system in which the desired result — a certain room temperature continuously maintained during certain hours — is achieved by use of a feedback control. Another familiar example is the steering of a car: even on an empty road, the driver's brain constantly monitors the relationship of the car to the sides of the road, and instructs the hands on the wheel to perform a continuous series of corrections to keep the car straight. This last example illustrates an important point about all systems, namely their tendency to go wrong. Philosophers and scientists call this tendency to run out of control 'entropy'. For our purposes this is the same as the old army expression SNAFU — Situation Normal, All Fouled Up. Although we are brought up generally to assume that things will be orderly, disorder is the normal state of nature, which has constantly to be fought against. The surprising thing is not that companies occasionally go bankrupt, but that more don't.

Any process such as a production line is capable of various outputs. The widgets might be just right, or they may be too hot or too cold, or too small or the

Systems analysis is the essential preliminary to effective use of computers R. & A. Bailey

wrong shape. Only some of these outputs will be desired. The more complex the system, the greater the possible number of outputs. Because quality management demands a very strict control on output, it puts a much greater potential strain on the processes than less controlled operations. In order to understand how this might affect the process, it is necessary to consider systems theory.

The basic elements of a system

Systems thinking attempts to see what all kinds of controlled operations have in common, so that the basic principles of control can be understood. Just as medical students discuss how diseases affect bodies in general, as well as looking at your body and mine, systems thinking tries to look for general principles. Four basic aspects common to all systems have been identified. These are:

1. *Structure*: How the organisation is built up from its parts, and how those parts react to each other — in the central heating system, the boiler, the radiators and the rest of the hardware.

2. *Content*: The work that the system does — i.e. the actual generation and distribution of heat.

3. *Communication*: The way that information about the system and the things that affect it are handled by the system — the wires that carry heat and switching instructions to and from the thermostat.

4. *Controls*: The making of plans, the setting of objectives and the monitoring of performance — Bridget's setting of the clock and the operation of the thermostat controls.

The process of control, which is the most important aspect of systems thinking, and the one that is most useful in quality work, is based on one simple concept — the feedback loop. If you touch a hot object, the nerves in your fingertips instantly relay the perceived temperature to your brain, which decides that some things are too hot to handle, and sends a message to the muscles of your fingers to remove themselves at once. The nerves have 'fed back' information to the brain, which compares the temperature recorded by the sensitive fingertips with what it regards as acceptable. If that level is exceeded, the brain relays a message to the muscular system to take action, and the finger is quickly removed from the source of heat. The similar, but less well known, concept of feed-forward describes the operation of a machine programmed in advance, such as a computer or a musical box.

The elements of the control function

Every system of control contains six elements. They are divided into two sets — the preparatory and the active. Before the control cycle can operate, each of the preparatory elements must be in place.

A. *Preparatory elements*
1. The control subject
2. The sensing mechanism, or how the control subject is to be monitored
3. The target, or standard to be achieved.

B. *Active elements*
4. Comparing actual with standard
5. Deciding on what adjustment to make to the process
6. Taking corrective action.

The control subject

A control subject is one attribute among several which is chosen to represent the state of the whole system. Thus the financial health of the company can be judged by return on investment, personal health by the thermometer, staff morale by absenteeism for trivial sickness and 'St Monday'. In each case a single indicator has been picked, which we know doesn't tell the whole story, but does reflect enough of it for us to work with. It is important to get the right indicator. We have seen in Chapter 10 how difficult service companies can find this. In

The control function

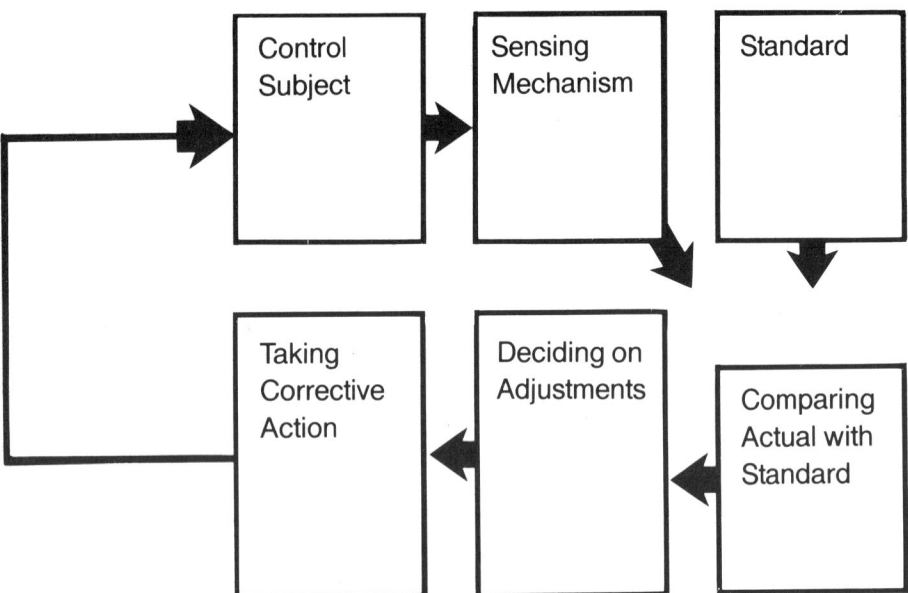

1. *The Basic Control Loop*
The sensing mechanism monitors the status of the control subject. This is then compared with the standard, and adjustments are decided on and then made. The new reading of the control subject by the sensing mechanism should be nearer the standard.

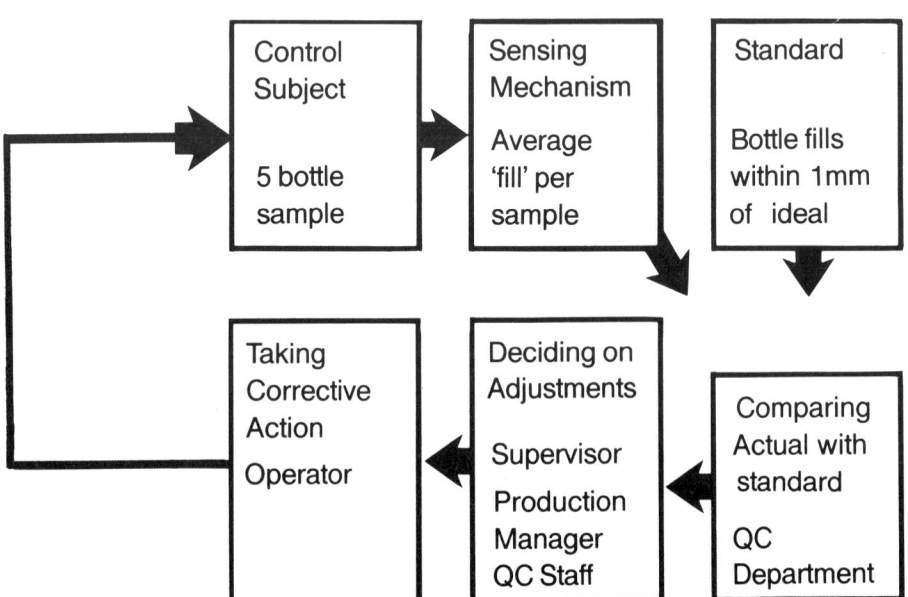

2. *The Control Loop in action: Controlling the 'fill' on a bottle filling line.*

manufacturing, the choice is likely to be easier, but there are still pitfalls to avoid. For instance, a company might decide to monitor production quality by the number of complaints from customers. The idea should work well enough for high priced items, but market research shows that people simply don't bother to complain about the quality of low priced items. They simply won't repeat the purchase. Because the quality monitor for low priced items won't signal any problems, the company risks losing sales as a result of bad quality.

Exhibit: The wrong control subject in dairy farming
Farmers are paid both for the quantity of the milk they sell to the creamery and its quality measured by butterfat level. In the long run, the only way to improve either of these factors from a given herd is by careful breeding. It is common in the dairy industry for a farmer to know how much milk his herd gives in total, and therefore what the average yield per milking cow is. The creamery tells him what the butterfat level as a whole is. Unfortunately these figures don't tell him whether it is, as it were, Daisy or Buttercup that produces the really high butterfat yield. As a result, when it comes to breeding or culling from his herd, he has only his intuition to rely on. This is one of the factors that has prevented the Irish dairy industry from producing levels of output rivalling European farmers. In system terms, they are using the wrong control subject, and consequently getting an inadequate result. Only when farmers can tell, cow by cow, what the yield in milk and butterfat levels is will they be able to develop their herds properly.

The control subject must be controllable. There is no point in blaming the purchasing manager for poor quality bought-in goods if the company insists on always buying inside the group. The weather, the £/$ exchange rate or the level of unemployment cannot by themselves become control factors, because no one in the company can do anything about them. If they are significant external influences, the control factor could be some kind of ratio connecting, say, sales levels to the dollar rate.

The sensing mechanism

Once you've decided what to control, the next problem is how to measure changes in the control subject. There are two aspects to this:
1. defining the unit of measure
2. designing a way of reading the current value.
Often the choice of unit and sensor is obvious. The heating system senses room temperature in degrees, the financial control system measures profits in money or return on capital, and so on. But even these well understood measures can get into difficulty. If a company estimates the value of its fixed assets on a conservative (i.e. undervalued) basis, and then compares its return on capital with other companies, the results will be comfortably high. Unfortunately they will be misleading, both to the company itself and to its shareholders. If the assets are correctly valued, a true measure of the company's performance will emerge. The difference can be startling, comparable to the difference between an

apparently generous wage rise, and that same wage rise with the rise in cost of living taken out.

If simple counting of results is not possible, there are various statistical ways of measuring output. Summaries, ratios, indexes, charts and rating scales are all familiar and are covered in detail in other chapters in this book. For our present purposes, it is enough to note that a measuring system must:
— *be countable*, that is it must use clear, separate and exact units, expressed as far as possible in the language of those doing the work;
— *reflect actual work done*, so that the unit measures the significant output — the thermostat should measure the room temperature, not the temperature of the water in the radiator;
— *be consistent* over time and in different places — the return on capital example above failed on this count.

The standard

The third preparatory element in designing a system is the standard by which the measurements of the control subject will be judged. Different places use different words to describe this concept: salespeople talk about quotas, accountants about budgets, production people use schedules; elsewhere the words 'aim', 'target', 'goal' and 'objective' are used. They all boil down to the outside judgement factor with which the measured output of the system is to be compared.

Nearly everywhere except in quality assurance the typical standard is based on previous results. Budgets are set up and compared with last year's figures, production controls are based on previous efforts, sportsmen compare their efforts with their 'previous best', or perhaps some record. This approach assumes that the factors that affected historical performance also affect future efforts. Most quality assurance programmes are directed by a concept of what performance should be, rather than what it has been. They do not therefore suffer from the problems of historical relevance. A standard should be:
— *Attainable*: There is no point in setting an impossibly high or impossibly precise standard; it should also be equitable and applicable to the conditions of work.
— *Consistent*: A major part of the science of metrology (see Chapter 18) is devoted to ensuring that standards of measurement remain consistent to remarkable degrees of accuracy. For most factory work, however, all that is necessary is to ensure that measuring devices are properly calibrated and agree with each other.
— *Economic*: The cost of setting and maintaining the standard should bear a reasonable relation to the activity measured.

The system is now prepared. We have decided on a unit to measure, and how, and the levels we wish to achieve. The next stage is to compare the measurement with the standard. This is the crucial moment of the cycle, because it starts the system on the path to self-correction.

Comparing actual with standard

The simple business of comparing budget with actual, out-turn with forecast, is the most basic activity of management. As long as the control subject and the unit of measurement have been well chosen, the only difficulty is in presenting the information so that crucial deviations from plan are quickly detected. In the early days of computers, it was common for executives to be given regular mountains of printout, sometimes amounting to 300 pages or more. These typically contained eight or ten units of information about several thousand items (products, sales accounts, budget lines, etc.). Some of those were off course: the problem was to find which. After a while it was discovered that the power of the computer could be harnessed to weed out automatically those elements which were on course, and merely present information on those not on target. Executives were presented with a much shorter list of items, each of which deserved attention. Considerable ingenuity has been deployed to enable the information required to be shown clearly: graphs, pie charts, standard report forms in all shapes and sizes, Gantt charts, Shewart charts, and Pert or critical path analysis formats.

Deciding on adjustments

The simpler the system, the easier it is to decide what to do about getting back on course. Steering a motor car normally requires only the slightest twist of the wheel to recover the correct direction. On an icy road, though, more subtle reactions are called for. In a complex system it may be very clear to the manager what is wrong, but not at all clear how to do anything about it. This is very often the case when more than one thing is going wrong at once. Doctors dread those complicated conditions where the patient is suffering from two or more diseases, perhaps compounded by the patient's old age or general debility, so that the normal method of curing the first disease would worsen the second. Before taking a decision, therefore, the manager should first of all investigate the problem. No system can ever be totally understood, so merely reacting to the control subject, by itself, could be disastrous. As we have already said, the control subject is only an indicator of what's going on.

Once the problem has been diagnosed, the manager must investigate and evaluate

---

Exhibit: Nappy training for executives: a routine for decision-making
This mnemonic gives the six steps that should be followed before taking a major course of action.
D – diagnose the problem
I – investigate resources available
A – appraise the situation
P – plan the solution
E – effect the actions required
R – review the results

the resources in hand and plan an active solution to the problem. Only when these stages are complete can the manager expect to put a sensible plan into action. A key part of the plan is the review of the effects of the plan. Has the action changed the value of the control subject in the desired direction?

Taking corrective action

Corrective action must be timely and appropriate. It must not be confused with the purpose of the system's activity, which is to continue to produce conforming products. The control and correction cycle, though very important, is only part of that. In fact if the system is running well, no corrective action will be required. Entropy (the SNAFU factor) will however prevent this happy state of affairs from lasting. For some managers action as such is attractive, and reassures them (quite wrongly) that they are doing their job. A. A. Milne described this kind of manager in action: 'It was going to be one of Rabbit's busy days. As soon as he woke up he felt important, as if everything depended on him. It was just the day for Organising Something, or for Writing a Notice Signed Rabbit ... It was a Captainish sort of day, when everybody said, "Yes, Rabbit" and "No, Rabbit" and waited until he had told them.'

Conclusion

The object of the quality disciplines is to produce consistently high relative product quality. This requires attention to detail and to the overall planning of the production process. To help managers plan, systems thinking has distilled the experience of many different kinds of control mechanisms, from the human nerve system to guided missiles. This general theory of control systems can then be applied in all sorts of different circumstances. The six elements of the control cycle are each essential, and can be discerned in any well designed quality control system.

Chapter 12

Motivation — Getting to Yes

WE all have jobs we like doing, and jobs we hate. The ones we like are done with love and attention, the others are scamped through. Inevitably, the results show which is which. Our attitude to the work critically affects the quality of the output. Therefore a successful quality plan, which aims to produce consistent quality work all the time, must pay close attention to motivation. By motivation we mean the various forces that urge workers to act as they do. High motivation tends to mean that the workers' objectives are similar to those of the company; low motivation means the opposite. This of course applies to office work and to quality assurance work as well as to production work. It is now well understood that in every area a highly motivated workforce is essential to high quality production. The question is — what are the factors that stimulate high motivation?

The need to motivate staff was not always as obvious as it now seems. In the early years of industrialisation, bosses regarded the machines as doing the real work and therefore providing the quality. The workers were mere adjuncts, whose task was to fetch and carry for the machines. They were dismissively called 'hands'. They were readily replaceable, whether by immigrant labour, by women or by children. As one group of Manchester manufacturers put it in 1854, the hands 'must remember that theirs is really a low species of skilled labour, and that there is none more easily acquired ... the master's machinery really plays a far more important part in the business of production'. The factory was rhapsodically described by one authority as 'a vast automaton, composed of various mechanical and intellectual organs, acting in uninterrupted concert for the production of a common object, all of them being subordinate to a self-regulating moving force' (Andrew Ure, *The Philosophy of Manufactures*, 1835). The workers were seen as inefficient, expensive and potentially rebellious cogs in this great machine. Attention to motivation was minimal; workers worked because they didn't want to starve — if they didn't work, or if they misbehaved, they were fined or sacked.

As factories became bigger and more mechanised at the end of the nineteenth century, an American engineer called F. W. Taylor introduced the practices of time and motion study to make them more efficient. Employees' activities were standardised and production stimulated by piece rate payment systems. The major premise of the theory was still that all employees were fundamentally the same, and increased efficiency could be achieved by analysing and then prescribing their every movement. Taylor studied the most efficient way of handling a shovel, for instance, and the method he chose was imposed on all workers. Money was treated as the sole key to motivation. The harder you worked, the more you got. 'Taylorism' had some early successes, but was quickly discovered to generate further distrust between workers and management. Despite this, some of Taylor's techniques are still widely used in modern time and motion studies.

In a long study in the 1920s and 1930s a researcher called Elton Mayo sought to develop Taylor's approach by identifying the ideal working conditions for a group of textile workers at the Hawthorne factory in New England. He reorganised them into smaller groups, and productivity went up; he changed the lighting, and productivity went up; he let them set their own work schedule, and productivity went up. To his surprise, however, it also went up when he reversed these conditions. Mayo was forced to conclude that what had motivated the workers to better productivity was the attention they were being paid as part of the research. Breakthrough! Workers were no longer to be viewed as 'hands', as cogs in a machine, but as people. The factory itself began to be seen as a social centre as well as a mere workplace. This more sophisticated view of what

Maslow's hierarchy of needs

was happening in factories necessitated something better than the old carrot-and-stick motivation system. Workers began to be seen as seeking both more complex extrinsic rewards (more pay, good social relations, status, etc.) as well as intrinsic rewards (pride of achievement, etc.).

Maslow's hierarchy of needs

The next development came from A. H. Maslow, whose theory was published in 1943. He believed that with their actions people attempt to satisfy five basic groups of needs (see diagram). A person was not likely to be interested in satisfying higher order needs until the lower had been sufficiently satisfied. Further, he thought that only unsatisfied needs actually motivated behaviour. After all, a person who has just eaten a good meal is unlikely to find the prospect of more food sufficient to tempt them to do anything. Management should therefore consider each of these needs while considering its personnel programme. The needs Maslow identified are:

— *Physiological needs*: The basic needs of survival such as food, shelter, air, warmth and rest.

— *Safety needs*: These are concerned with protection. They include the desire for job security, health, pension, savings and insurance. Physiological and safety needs are the basic needs which must be satisfied before the higher needs can be met.

— *Social needs*: The requirements for acceptance by others, for giving and receiving friendship and affection and for belonging.

— *Ego needs*: Including the need for self-esteem and the need for the esteem of others. These two needs are closely connected, since the response from others bolsters self-esteem and vice versa.

— *Self-fulfilment needs*: These are the needs related to creativity, to power-seeking, to the need to develop one's human potential.

Maslow's theory usefully stresses the fact that everyone has a variety of needs. Without denying the importance of money and job security, it points out the motivating strength of the creative and self-fulfilling urges. The quality concept rests heavily on these needs, as did the old-fashioned craftsmen. Maslow's theory has stimulated a great deal of work on the same lines. In practice, however, there is no evidence that the actual hierarchy of needs has universal validity. The relevance of the five groups differs from person to person. There is also no evidence for the view that a satisfied need does not motivate behaviour.

Theory X and Theory Y

Taking Maslow's theory for granted, Douglas McGregor discussed the kind of management style appropriate in an economy where most of the basic needs of the workers were already met. He distinguished the old style of management, which was based on authoritarian, military models, called Theory X, from the new, liberal Theory Y. These managerial assumptions worked as self-fulfilling prophecies. If managers treated the workers in a 'Theory X' fashion, that was how they behaved, and equally with 'Theory Y' style management. Each style was characterised by an underlying set of beliefs.

1. *Theory X*: Workers are seen as naturally lazy, uncooperative, passive, irresponsible, uncreative and motivated only by money. As a result, the only way managers can operate is by persuading, coercing, rewarding and (frequently) punishing workers. People cannot be trusted to work effectively without active and constant supervision. This is the environment that produces low quality products while a large proportion of quality costs is spent on inspection and reworking. In Maslow's terms, Theory X managers fail by concentrating totally on lower order needs. Lower order needs are relatively soon satisfied, so their motivational power is slight. By ignoring the higher order needs, these managers actually kill the most powerful sources of motivation. As a result they have to resort to punishments such as unpleasant work allocation, loss of pay or even sacking. This amounts to reviving the lower order needs relating to survival in the worker, who will then be motivated to satisfy them.

2. *Theory Y*: In this theory, work is a natural activity and should be a source of satisfaction. All workers are capable of self-disciplined and creative work. Satisfaction of workers' higher needs can be made to be consistent with the company's objectives, among other things by proper quality planning. The job of management is to harness and channel this inherent intelligence and creativity to the right ends. Even people who seek job satisfaction from the work itself can quickly be deterred by Theory X management attitudes. They then look for the other benefits of the work environment, and treat the actual job as a means to those ends. This atmosphere does not develop the workers' self-esteem, nor does it produce high quality output.

Maintenance and motivating factors

A scheme that took the same idea of a set of needs was produced in 1959 by Dr Frederick Herzberg. He divided needs into maintenance factors (roughly the same as the physiological and safety factors in Maslow's hierarchy) and motivating factors. Typical motivators are advancement, recognition, responsibility, challenge, status. Motivating factors tended to push workers in a 'Theory Y' direction. 'When they felt positive about their jobs, they put more care, imagination and craftsmanship into their work; when they felt negative, they were not necessarily careless, but neither did they worry about the fine details,' Herzberg wrote.

A study at the US Texas Instruments plant in the 1960s suggested that this division could also be used to distinguish two groups of workers. *Motivation seekers*,

Two SI employees with their slogan which has been incorporated into all product packages
System Industries

are driven by the factors relating to the task, and tend to have higher tolerance for a poor environment. *Maintenance seekers,* on the other hand, regard the basic factors as more important. They are typically outer-directed and show little interest in the task itself. Taylor's scientific management approach had in effect imposed maintenance seeking attitudes on the workers, which is why it failed as a theory of motivation.

Process models of motivation

The theories of motivation described above attempted to discover what elements in the work situation actually motivated people. They are called *content theories.* Another approach is to treat the individual as performing a quick mental

benefit/loss calculation when presented with a task. These theories are called *process models*. The individual is seen as considering the advantages and disadvantages to him or her of:
1. the rewards of successful outcome of the task,
2. the likelihood of that outcome if everything goes well, and
3. the chance of everything going well.

The result of this mental calculation dictates the amount of effort put into the job. This theory emphasised the fact that what motivates people is how *they* see things, not some objective reality.

These theories show the person making a series of rational choices, based on their values and the information to hand. If a person believes that rewards come from not being found making mistakes, rather than actual quality of performance, the rational decision is to minimise errors rather than to maximise quality. Because the values of one individual are different to those of another, the same situation can produce quite different, but equally rational, individual attitudes.

The sociology of the workplace

The theories of individual motivation are difficult sometimes to relate to the complexities of the shop floor. Everyone in the company brings their own educational, social, sexual, religious and political attitudes into the workplace. For instance, people's attitudes to work are strongly influenced by their social and economic background. Middle class people tend to start work with high expectations, whereas working class people generally expect it to be burdensome, restrictive and monotonous. Research shows, however, that both of these culturally induced attitudes can be broken down quickly.

Elton Mayo discovered that the workplace is not merely, or for many people mainly, a place of work. As one pools winner put it, having returned to work after a very big win, 'When I won I had a really good time for a couple of months, and we bought our house, but I missed my mates ... it's no fun being on your own all day.' Certainly money is important, but other aspects of the work environment can be more so. However, many of the benefits of these aspects, such as friendly relations with fellow workers, job security, etc., can easily be achieved without making any internal commitment. If workers' self-esteem is not affected in some way by the quality of the work, quality will sink. In fact people prefer to be proud of what they do. But generating this pride is not done by management speeches in the canteen. It must become a real part of the corporate culture.

A corporate culture is the set of understandings and attitudes shared (more or less) by everyone in the company. It is positive or negative depending on the values it encourages. Negative corporate cultures are engrossed in office politics, and tend to concentrate on the figures rather than the product. If the MD only wants to know about productivity and not at all about quality standards, these values quickly become part of the corporate culture. The touchstone of corporate culture is the way in which managers lead by changing people's actions and ideas.

Getting and giving orders

An important part of workplace sociology is the system for influencing peoples' actions. In the Bible, the centurion said 'Go,' and the soldier went. Modern managers usually use rather more subtle techniques. Conscious of motivation theories, they seek if possible an inner assent from the workers. As a recent book put it, they are interested in 'Getting to Yes'. However not all management techniques for direction are so positive. In fact the basic styles can be divided into two sets: collaborative techniques and directive techniques.
Collaborative tactics
1. Reason — 'It's the best thing to do.'
2. Friendliness — 'Do it for me.'
3. Coalition — the 'We're in this together' approach.
4. Bargaining — 'Do this and I'll make it up to you.'
Directive tactics
5. Assertiveness — 'Do it.'

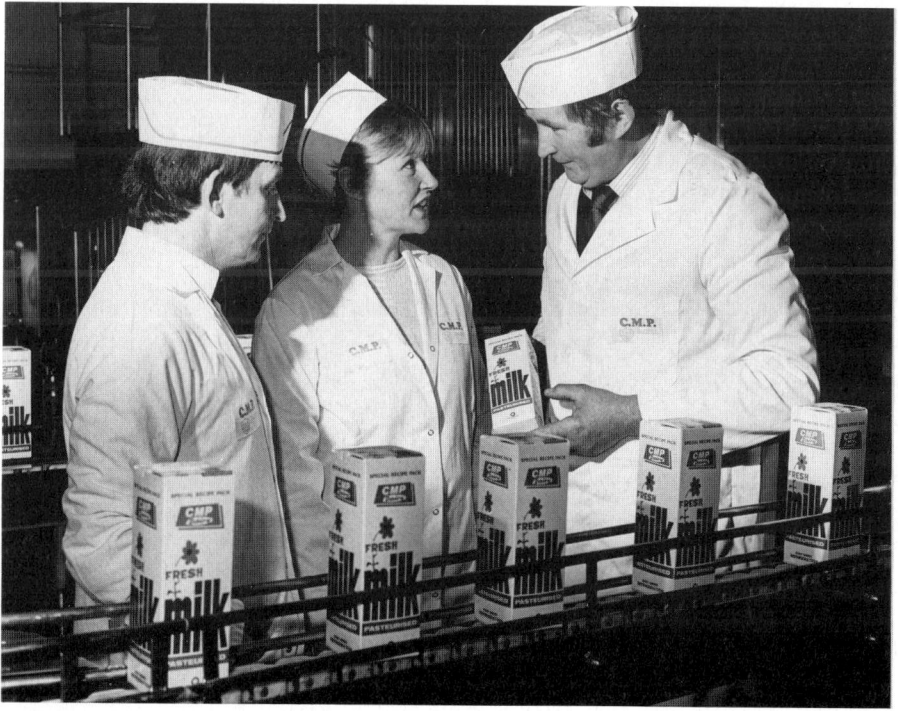

Supervisor discussing seal quality of cartons with operators   *CMP Dairy*

6. Higher Authority — 'The MD says you're to do it.'
7. Sanctions — 'Do it or you'll be sacked.'

The choice of tactic is dictated by the circumstances. If a manager wants something done that is a little outside the direct command line, such as extra overtime, or unusual work, collaborative tactics are likely to be used. In general, managers decide on which tactic, or combination of tactics, to use by the following general rules.

1. *Power*: The greater the direct power of the manager, the wider variety employed.

2. *Pecking order*: Managers use various tactics (ranging from friendliness to bargaining and higher authority) with their subordinates but have to rely mostly on reason with superiors. This is because their power base is smaller, and so the range of tactics available is less.

3. *Power distance*: The greater the power distance between the manager and the subordinate, the more likely the manager is to use directive tactics.

4. *Preference for collaboration*: If there is a reasonable chance of success managers prefer to use collaborative tactics. The more positive the corporate culture, the more likely it is that collaborative tactics will be used, because the right values are shared.

Whichever tactics are preferred, consistency is important. A study of supervisors in a General Electric plant suggests that workforce motivation is higher under bosses with consistent use of tactics, rather than those who veer widely and unpredictably from friendliness to sanctions. Clearly the workforce prefer predictability.

Faced with this barrage of management technique, the worker has three choices: reject the attempt at influence, ignore it, or accept it. There are three ways to accept a direction.

1. *Compliance*: The person agrees to the request because it is more worth while to do so than not. Force, rules, procedure, and custom are all involved, but in the end compliance always implies 'has to'. The recipients concur with the request, but grudgingly, since it denies their choice in the matter. From the manager's point of view compliance is quick; no time is spent in explanations. On the other hand mere compliance requires constant checking and back-up power to enforce it.

2. *Identification*: The person adopts the proposal because of admiration or identification with the source. This is the way a charismatic leader operates. It works for the leader, or for the group who have identified with a task, but it is inefficient for the company as a whole. If the leader moves on, or the task

is no longer necessary, the group cohesion is lost. Even while the influence is operating, the nature of the relationship usually prevents subordinates from exercising intelligent choices over their activities. Companies, armies and churches have all found that charismatic 'commando' leaders are effective in the short run, but are ultimately disruptive.

3. *Internalisation*: In this mode of acceptance the person adopts the new idea or behaviour pattern as their own. It will be acted on without stimulus (as in identification) or pressure (as in compliance). The change will be self-maintaining, and independent of the original source. The influenced person will tend to believe that the change was their own idea. This can be difficult for managers, who, like everyone else, prefer to receive the credit for their ideas.

Incompetence

Another vital aspect of corporate culture is the staff's confidence in the competence of management. In the First World War the managers (generals) sent wave after wave of men charging against entrenched machine-gun positions. This tactic, despite strong evidence of failure, was hardly changed from the beginning to the end of the war. The German generals understandably referred to the British army as 'lions, led by donkeys'. Whatever tactics the manager uses to influence people, nothing demotivates them more quickly than apparent incompetence. This doesn't mean that managers have to be right all the time. That is impossible. But it is important to appear rational. The instructions must be seen to be consistently related to a known set of objectives, to well known procedures and to changing circumstances. If the managers are seen as incompetent, the staff suffer from insecurity. Ultimately they fear for their jobs, immediately they suffer a loss of the crucial motivator, self-esteem. If the section as seen as so unimportant that incompetent supervision doesn't matter, morale will quickly suffer.

The practical lessons of motivation theory

1. *Charity of discourse*: All modern theories of motivation stress the importance of workers and management treating each other as rational human beings. The concept of the 'charity of discourse' develops this by suggesting that, unless there is overwhelming evidence to the contrary, the other side should be assumed to be truthful and seeking a fair resolution. However obstructive and bloody-minded the other side appears to be, they must be assumed to have good reasons, from their point of view, for what they do. It is therefore in the motivator's interest to understand that set of reasons.

2. *Feedback*: Once there is reasonable agreement, it is the motivator's job to

reinforce that. This is done much more effectively by praising (and rewarding) good work than by punishing bad. The human feedback provided by opportune praise can work wonders. Workers' self-esteem is an important area of motivation. Everyone after all likes to be approved of. Unfortunately for them and their staff, some managers find it very difficult to be generous with praise. This attitude makes high motivation hard to achieve. It often reflects an insecure unwillingness on the manager's part to come down from the position of supreme authority.

3. *Participation*: Employees should be allowed to participate in all decisions as far as possible. Participation is another technique that requires a sophisticated attitude from the motivator. Inexperienced, nervous or incompetent managers are often reluctant to allow subordinates to take part in decision-making. Yet clearly it is the quickest route to that ideal mode of acceptance, internalisation. After all, if one has taken part in the making of a decision, it is much harder to reject it. The other motivational advantages, such as recognition, affiliation and acceptance, are also important.

4. *Job enrichment*: Nobody can expect workers doing boring jobs day in, day out to be highly motivated. Managers must therefore at least rotate the staff to provide variety, and at best develop job enrichment schemes. Job enrichment involves such things as planning more variety into the task, giving workers a chance to choose their work plans and methods, improving the sense of team work in the group and in the plant generally, setting quality and other goals, providing feedback on performance, and improving the environment. An important part of the quality plan is motivating the workers to accept it: their self-esteem can be stimulated as they produce a better and better product. The achievement of the quality targets then becomes a motivating factor in itself.

5. *Money*: Simple-minded motivators have used a combination of money incentives and threats of the sack as their sole motivation armoury for generations. Psychologists believe however that after a certain level has been passed it has a rapidly diminishing effect as a motivation force. Of course its effects are confused by the fact that it can buy such things as prestige, which is a significant ego need for some people. A major problem with using money as a motivator is caused by the way the tax system works. Money works best as a motivator if the sum offered is large in proportion to present earnings. Unfortunately, if the recipient is on high marginal tax rates, it can be very expensive for a company to make a sufficiently attractive after-tax gesture.

Chapter 13

Putting your point effectively

IT is a damp day in the west of Ireland. Rain seeping everywhere. Into boots, into the tops of macs — everything gets wet. An English tourist on his daily walk meets a local farmer. He touches his cap with his stick. 'Morning Sean,' he says. 'Damp again today.' The farmer agrees: 'It's a soft day, thank God.'

An ordinary enough encounter, but still puzzling. Why for instance are these two grown men standing in the rain telling each other that it is raining? And why does one call it 'damp' and the other 'soft', when the rain is coming down in buckets? Why does the farmer bring God into the matter? And how does the farmer interpret the waving of the stick?

Human communication is a complicated process, but a vital one. As the exchange above shows, it is about much more than mere exchange of facts. Chapter 8 stressed the importance of information by saying that a company without good information systems is like a car without petrol. Communication is the name for the various techniques for conveying information. Very often the contextual and non-verbal information says more than the verbal. For instance, a negotiator who sits with his arms folded across his chest is probably rejecting your point of view, whatever he may say verbally.

Many people think that the method of communication we normally use profoundly affects our view of the world. As the psychologist Abraham Maslow put it: 'If all you have is a hammer, every problem is a nail.' The more one is limited to a particular medium, the more one becomes unable to envisage the world except in those terms. In business, scientifically trained people often have great difficulty in explaining their ideas to financial people. Thus an engineer may be able to describe in detail what is going wrong on the shop floor, but unless the point is expressed in financial terms, the accountant won't be able to grasp the significance of what the engineer is saying. They are in fact talking a different language, and expressing a different view of the world.

Types of communication

Communication is normally divided into four (or, if you include telepathy, five) categories. These are verbal and non-verbal communication, mass communication and humour. Humour is in a special category because it is the only form of communication in which highly complex verbal messages (jokes) are responded

Quality in Practice

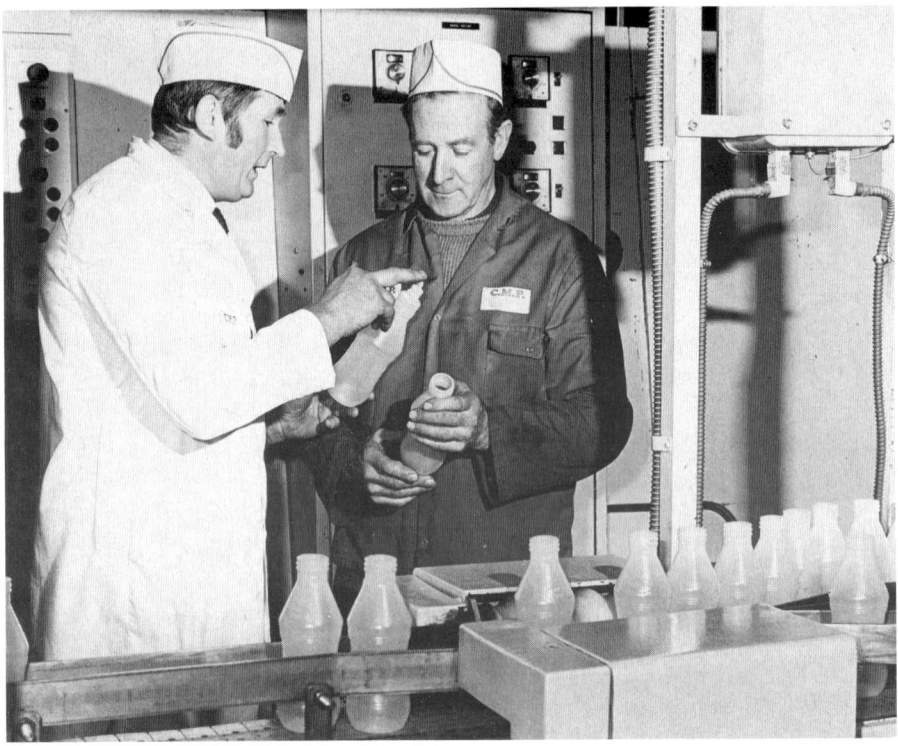

Practical communication on the line   *CMP Dairy*

to by a reflex physical action of fifteen muscles of the face (laughter). It spans the mind/body combination in a unique way. Unfortunately humour is not usually part of the quality management syllabus.

1. *Non-verbal communication*: Academics distinguish various types of non-verbal communication by the density of meaning they carry. The very simplest are *signals*, such as a tap on the window, which is hardly more than an interruption. Though they carry no meaning in themselves, they can be used for instance to alert operators that some (unspecified) thing has gone wrong on the line. Bells, alarms, hooters and flashing lights can all serve this purpose and others.

Signs are more explicit. They carry an explicit verbal or drawn message, such as No Smoking, Danger or Quarantine Stock. A fully worked out and understood set of plant signs and signals, as part of the quality plan, also carries a message to the workforce that quality matters at every level. Visual aids used in the presentation of data are also signs. Graphs, histograms, pie-charts and other visual methods often convey the information much more quickly than any amount of verbal analysis.

Carrying a richer burden of meaning still are *symbols and icons*. These are

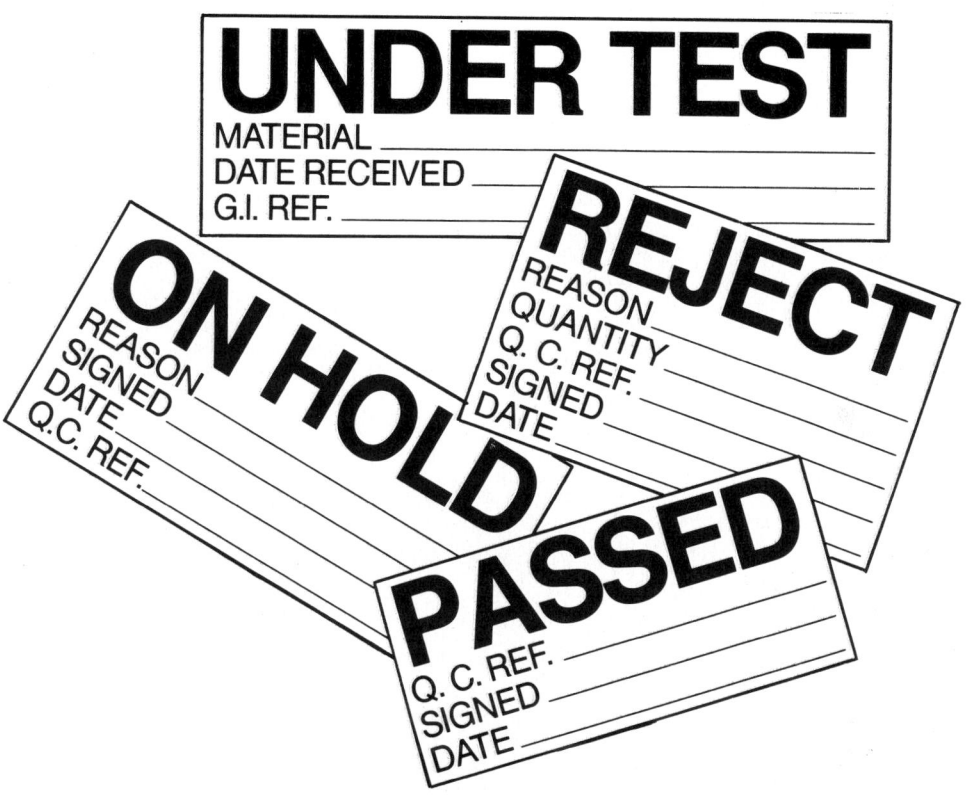

A clear set of QA goods inwards labels   R. & A. Bailey

shorthand terms and images that carry an especially powerful weight of meaning for a particular group. The Union Jack, the Statue of Liberty, the GPO and the Siege of Londonderry are all classic religious or political icons. Membership of a group implies acceptance of the enhanced meaning of the icon. In company terms icons and symbols make up the corporate culture which is an essential part of the quality plan. Nicknames, stories about the plant in the past, about the bosses, about strokes, triumphs and disasters tell newcomers what the company *really* values. The executive washroom and canteen are classic symbols, often carrying more information about the personnel policy of the company than any amount of well meaning memos.

Finally, there is the repertoire of gestures and physical relationships that makes up *body language*. In human relations, this can be of immense importance. Scene one: You are invited into the boss's office. She remains seated as you come in, continues writing, and makes no attempt to offer you a chair. Scene two: You are invited into the boss's office, but this time she comes from behind her desk, invites you to sit down, and sits beside you. Not a word has been spoken, yet we all know which means promotion and which doesn't.

2. *Verbal communication*: Most people think of communication as being mainly about words. In fact non-verbal messages can be just as powerful and important. However, words are able to carry precise, complicated or elaborate messages. Journalists often say, 'One picture is worth a thousand words.' What they mean is that a picture can convey some of the non-verbal messages about an incident that are so important. But for conveying precise information a picture can contain too much distracting and irrelevant information. No picture of the Shannon can evoke in the viewer's mind as explicit and general an image as that one word. Even a map is useless without placenames.

In trying to convey these verbal messages, we all use several different sets of vocabulary. The words and even the grammar that are used at home are often different from those used at work. In the workplace people use different jargon with different colleagues. Slang and jargon arise naturally as part of the corporate culture. The fact that it is often barely understandable to outsiders, even those in the same trade, is inefficient, but generates an obscure satisfaction. Part of the purpose of slang is to express group solidarity.

One highly specialised form of language is computer programming language. A great deal of recent work on language has been done as part of an attempt to enable computers to use natural language. Unfortunately, the research so far has simply revealed how much more complicated human language is than was once thought. The problems computers have with translating from one language to another are well known. One machine translation of a technical manual inserted 'water sheep' for 'hydraulic ram' throughout. The difficulty is that *all* human communication takes place in, and refers to, a context. Misunderstand or remove the context, and much of the meaning is removed. Take the two statements 'Helmets must be worn' and 'Dogs must be carried on the escalator.' Grammatically identical, yet no one would fail to mount an escalator for lack of a dog to carry. Programmers still have great difficulty in defining and then specifying the context to computers.

3. *Mass communication*: Verbal and non-verbal communication has always been part of human culture. The use of technology to distribute messages to thousands of people at once is effectively an invention of this century. Mass circulation newspapers, radio and, above all, television now provide much of the source and matter of daily conversation. Gossip is as much about The Riordans or Dallas as about the neighbours. The simple act of watching and later discussing a US series involves at least some acceptance of a system of values and a vocabulary that would have horrified previous generations.

The mass media are important to managers because they form the context in which normal private and company messages are read. We are all constantly exposed to the mass media. Most people watch at least nineteen hours of television a week (i.e. more than half the number of hours spent working) not to mention time spent listening to the radio. Research in the US states that the average American is exposed to 1,600 advertisements every day, notices about 1,200 and actually responds to 12. We have become sophisticated at *not* attending to messages we don't want to hear. At the simplest level, this means that a notice written for those who read the *Star* and the *Sunday World* has to be expressed in similar language if it is to be fully understood. At a more complex level, the mass media vividly present values and habits (particularly behaviours and attitudes) that may be quite unsuitable to Irish circumstances. If workers on strike are exposed nightly to news bulletins about violence on the picket line in Wapping and other places, they will be inclined to think that that is how picket lines are properly conducted, even in Ballydehob.

Written communication

For most people the most difficult form of communication is the written word. During a conversation we can see if the other person has understood the point by their facial expressions and head movements. It is part of the polite listener's role to provide this regular affirmation. If necessary, we go over the ground again. It is not essential to worry about precise grammar or vocabulary or punctuation. In writing, all these things become important. Small slips can significantly and irreparably change meanings: a little, wanted baby is not at all the same as a little-wanted baby. The rest of this chapter will therefore concentrate on the problems of writing English.

Written language attempts to transfer a message from one mind to another by means of words, arranged in certain order, and laid out on the page in a special way. The words, the syntax and the appearance are all important in helping the reader understand the message. But before they come into play, the writer must decide what is to be said.

Everyone is familiar with the idea that love stories have a standard pattern. Boy meets girl, they fall in love, complications arise, they are resolved, the pair marry. The names and the settings change, but the pattern carries on. This pattern can be further generalised to other kinds of stories. In fact George Bernard Shaw once said that there were only nineteen different plots in the whole of literature (he also claimed, intriguingly, that only six of them had ever been used). In the same way, much ordinary company writing (memos, letters, reports, standards, etc.) can be fitted into a pattern. In virtually every case the text answers four basic questions:
— what is the situation?
— what is the problem?

— what is the solution?
— how well does it work?

This is the *Situation — Problem — Solution — Evaluation* pattern. It can be used as the framework for any piece of writing.

The framework also helps the writer with the perennial problem of selection. The first element, the situation, needs to be described only as far as it relates to the problem. Every statement is a selection from the wide range of things that might possibly be said. In a memo about increasing customer complaints, the fact that the complaints came by post or by phone is irrelevant to the problem or the solution. The complaints may be all from women, for instance, or all arrive in the morning, or all be from Cork, or all be very politely expressed, or all written on lined paper; any of these could be relevant or irrelevant.

The situation leads directly on to the problem. From one point of view the problem might be a lot of dissatisfied customers, from another an increased workload in the customer relations section, from another a design defect in the product. Each of these is a problem that begs a solution. Just as the problem evolves from the situation, the type of solution emerges as a result of the statement of the problem. Thus if the problem is seen as a design fault, providing extra staff in customer relations is not a solution.

Finally, the memo should be rounded off with some estimate of the effects of the solution. This might be based on experience, on assessment, on expert or amateur opinion. It rounds off the statement, though it might well point out that the solution proposed or adopted may or does give rise to further problems. The cycle then starts at a new situation/problem point.

The pattern can commonly be seen in condensed versions. Most typically condensed is the Situation aspect. The context can often be assumed. For instance a road sign reading 'Stop — Accident Ahead' is a problem and a solution. The situation hardly needs to be mentioned. 'Heroin kills' is a problem, with an implicit solution. Once again the situation is implicit: it doesn't kill people who don't take it. Other familiar condensed structures are Solution-Evaluation, for instance the slogan, 'Quality is Free.' The kind of essay we did at school on 'A Day at the Seaside' is a condensed Situation-Evaluation structure. If in doubt, the basic pattern should be stuck to, even if the situation is mentioned only in a single sentence. Before writing a memo or a report it can be useful to write down all the bits of information in four columns, headed Situation, Problem, Solution, Evaluation. This helps to clarify the mind on the topic and to get the pieces on paper in the right order. The elements themselves should be kept in the natural order.

Using language

Once we've decided what to say, the question is how to say it. It is important to distinguish between the stage of deciding what you want to say, and the stage of choosing the words. The first choice of words is not always best. The same glass can be described as half full or half empty; the same star (Venus) can be the evening star or the morning star. The right choice can make a major difference both to meaning and readability.

1. *How readable is your prose?* The readability of English prose can be roughly measured by the Fog Index, devised by the American teacher of clear writing Robert Gunning. The formula is based on two of the factors that make reading difficult: long sentences and long words. For long pieces of prose, the index can be calculated from hundred-word samples. The Fog Index formula adds the average number of words per sentence to the percentage of long words in the passage. The result is then multiplied by 0.4 which brings the index number down to a manageable size. Long words are defined as words of three or more syllables. Exceptions include proper names, words that are combinations of easy words (such as everything, bookkeeper, manpower), and verbs made by adding -ing, -ed, or -es to two syllable words.

Exhibit: Calculating the Fog Index
The general formula for the Fog Index is:

$$FI = \left(\frac{a}{b}\right) + \left(\frac{c}{a} \times 100\right) \times 0.4$$

where a = number of words in chosen passage
b = number of sentences
c = number of long words

The first nineteen sentences in this chapter contain 217 words. The average number of words per sentence is 11.4. There are eighteen long words, giving a hard words proportion of 8.3 per cent. The formula therefore reads:

$((217/19) + (18/217 \times 100)) \times 0.4 = 7.88$

The long words were: ordinary, encounter, interpret, communication (twice), complicated, importance, company, various, probably, information (four times), contextual, non-verbal, negotiator, verbally. If I had wanted to reduce the Fog Index, I could have changed 'encounter' to 'meeting', 'complicated' to 'complex' and so on. This however would be misusing the index. It is a measure of readability, not a guide to writing. It is a control subject which reveals the state of the writing.

In fact this Fog Index level is untypically low, because the first 100 words are mainly dialogue and score as low as 4.4. The next 100 words score rather higher. The originator of the index, Robert Gunning, advised 'Anyone who writes with a Fog Index of more than 12 is putting his communication under a handicap, and a needless handicap at that. For almost anything can be written within the easy reading range.' He reports that the *Readers' Digest* has a Fog Index of 9, and *Time* magazine 10. Index level 13 is the level of a first year university student.

The Fog Index is quite easy to calculate, but it does ignore some factors that make for hard reading. Sentences with passive verbs, heavy use of auxiliary verbs, abstract nouns, and many subsidiary clauses are all relatively hard. Despite these limitations, the Fog Index is a simple tool for testing readability, and should be regularly used to monitor information quality standards in memos, manuals and other vital pieces of company information.

2. *The basic sentence*: Rule one of readable English is to follow the standard sentence form unless there is a good reason not to do so. Nothing causes more difficulty than twisted and misformed sentences. The basic form of the English sentence is simple. It runs Subject — Verb — Object — Auxiliary. Sentences such as:
— John rode the elephant
— Quality assurance will inspect all new equipment
— Every product has quality costs

all fit this basic mould. A wrong sentence order has the effect of forcing one word to your attention out of its place, as in this sentence: It is a fact that the important points of a sentence are frequently in a minor position. The significant part of the sentence is pushed into a minor 'that ...' clause, while the mere comment 'it is a fact' becomes the key part of the statement. If this is done accidentally, it forces the reader to look for reasons for this emphasis that aren't there. Any departure from the standard order should be done self-consciously, for a specific reason. The typical reason is a desire to stress one particular aspect of the sentence. For example:
— On the floor lay the weapon, a heavy poker.

has artfully pushed the stress on to the poker and its position on the floor by moving it from its normal position before the verb. (The standard sentence would read: The weapon, a heavy poker, lay on the floor.) Unfortunately, just as a lie exploits the fact that most of us tell the truth most of the time, unless the normal sentence is adhered to, the dramatic effect of a change of order is lost.

The second basic rule is that any adjectives, adverbs or other descriptive words or phrases should be as near as possible to the word they modify (technically called the 'head word'). Adjectives generally precede the head word, and adverbs follow. Once again this principle can be flouted if the author *deliberately* wishes to stress something.

Rule three of writing is to be as clear and interesting as possible. For instance, a regular source of reading difficulty is the constant use of weak auxiliary verbs (such as 'will', 'must', 'may', 'could', 'is') in place of lexical verbs (ordinary verbs such as 'measure', 'run', 'tease', 'produce'). Some authors have trouble in committing themselves to anything. Their statements are hedged around with words like 'probably', 'relatively', 'nearly', and 'possibly', until the reader stumbles in a fog of relativities. Another aspect of this shyness is the constant use of passive verbs. With phrases such as 'the level was measured', the 'production line was stopped', or 'the stock was counted', it is as if the process were performed by ghostly hands, or by people who prefer not to admit they exist.

Rule four is that the writer should never be afraid to revise and rewrite. The draft of an important note should first be tested for Fog Index level. Then it should be assessed according to the information quality aspects discussed in Chapter 8. Does it say what you meant to say? In what context will it be read? English is a slippery and difficult language, which needs to be worked at. The first draft is rarely adequate. English grammar is full of mysteries — for instance the secret set of rules that dictate the order of adjectives in front of the head word. Phrases such as 'a black big hunting dog' or 'a Norwegian young sailor' are obviously wrong, but it is not clear why. Another oddity is the way the metaphor of 'up and down' permeates the language: happy is up, sad is down; in control is up (on top), subordinate is down; rational is up, emotional is down ('the discussion fell to the emotional level').

Rule five of writing is to take care in choosing words. On that no more can be said than this, written in 1906:

> Anyone who wishes to become a good writer should endeavour, before he allows himself to be tempted by the more showy qualities, to be direct, simple, brief, vigorous and lucid. This general principle may be translated into practical rules in the domain of vocabulary as follows:
> — Prefer the familiar word to the far-fetched.
> — Prefer the concrete word to the abstract.
> — Prefer the single word to the circumlocution.
> — Prefer the short word to the long.
> — Prefer the Saxon (English based) word to the Romance (Latin or Greek based).

Conclusion

Techniques of verbal and non-verbal communication are important. We have seen that information is a critical area in the quality plan. In devising the quality plan great attention should be given to the communication aspects. The signs and symbols around the factory should be carefully devised. The new icons that reflect the quality orientation of the company should be considered. But most important of all, the quality of verbal information should be worked at. Specifically this involves following the five cardinal rules of prose writing:
1. Always follow the standard sentence form of Subject — Verb — Object — Auxiliary unless it is absolutely necessary not to.
2. Keep modifying words and phrases near the words they modify.
3. Be as direct as you can; don't hedge your statements with passive constructions or relativities.
4. Rewrite.
5. Choose your words carefully.

Chapter 14

Training for quality

THE achievement of quality requires the company to harness the energies and materials of nature, as crafted by technology, and the creativity and skills of human beings. These skills do not occur spontaneously. They have to be learnt and then developed. In the past learning was a once-for-all activity. The young apprentice left school having learnt to read and write, and then spent seven years with various masters learning the skills of the trade. He was then a journeyman or a craftsman, and able to work on his own. His skill and experience thereafter developed by doing the job. Nowadays the technical content of jobs changes so rapidly that any employee (management or shop floor) who doesn't regularly update will be only half as useful after ten years. Recognising this has led quality conscious companies to put enormous emphasis on training. Training programmes are now a regular part of quality plans and are no longer reserved for introducing new machinery or new technology, or for combating the effects of high staff turnover.

Training may be aimed at improving the recipient's knowledge, or skill, or at changing attitudes. Generally any specific programme partakes of all three, but they are separate aspects of the training task. If new machinery is being introduced, the operator will first need to gain knowledge about the machine (how it works, what it can do, what the dangers are, etc.), then acquire the skill to work the machine and finally the motivation to get the most out of it. All training should be aimed at gaining specific results, in terms of productivity, quality, motivation, etc. There are four types of training that companies normally engage in, which should all be part of the training plan.

1. *Induction training*: This is generally aimed at the new recruit, and covers general aspects of the company, such as history and objectives, as well as more specific aspects of the job.

2. *Retraining programmes*: This part of the training plan involves covering some of the same ground as the induction programme, and generally ensuring that the standards set then are maintained.

3. *Upgrading programmes*: As technology develops or markets change, staff have to be trained to make the best use of the new machinery.

4. *Development programmes*: These are aimed at staff who are about to be promoted to new responsibilities.

Exhibit: The first day on the job at Ballyfree Farms, Co. Wicklow
The first day on the new job is probably the most important day in the employee's career with the company. First impressions can often be lasting impressions. After the employee's supervisor has gone through the usual PRSI documentation, medical certification, break times, plant rules, protective clothing, etc., the employee is introduced to the microbiology laboratory supervisor. The employee spends a portion of the first day in the laboratory. During this time a basic knowledge of what bacteria are, how they multiply, the impact of temperatures and the diseases they cause is grasped; in other words, a basic knowledge of how to grow them, how to kill them, the effect of heat and cold, the effect of cleaning procedures and the cleaning principles involved are outlined. The significance of hand-washing, protective clothing procedures and the health monitoring programme are described in detail. The employee then begins learning the components of the job under the guidance of the area supervisor/instructor.

In addition to the individual training programme the company runs an in-house group training programme on quality and hygiene during the winter, mainly concentrated into January and February. It is organised by the quality control manager and includes inputs from individual managers, other members of staff, and technical representatives from our customers and suppliers. There is also a programme of visits to supermarkets. One of the marketing personnel takes a group of five or six people from the plant and tours four or five supermarket outlets in Dublin. This gives the staff a chance to look at the product from a presentational, quality and image point of view, and also to compare it with other products.
Source: Dr Pat Griffin, Processing Director, Ballyfree Farms

Helping the learning process

A training need exists because there is a gap between what the employee can do and what the company situation requires. In many cases that gap can be filled by training. This might be training in skill, by teaching new or better techniques, or the development of knowledge. Training programmes deal with adults who must be approached by different techniques to those used in school. Apart from anything else, most adults are quite unused to sitting for long hours being lectured to, and would resent any attempt to recreate a classroom atmosphere. Very often they enter the learning situation with little confidence in themselves, and with a fear of making mistakes, or exposing their incapacity. There are six key elements in developing a successful learning situation:

1. *Motivation*: People who don't want to learn, won't. Some reasons why they might want to include: curiosity, self-preservation, self-betterment, and a desire for more responsibility. The benefits to the students should be stressed all the time. Constant feedback on performance enables an adult to monitor his or her own performance, and to maintain motivation.

2. *Past experience*: Successful learning requires that new knowledge be related to old. If the new knowledge or skill develops naturally out of the old, it is easier for the trainee to see its relevance to real life. Also, something that doesn't have a clear relation to what is already known or believed will quickly be forgotten.

3. *Easily absorbed units*: New knowledge is best taken in small doses, in logical order, so that one small step forward follows another. This technique enables the trainee to digest the new information easily, and allows the trainer to check that each stage is accepted before moving on to the next.

4. *Good learning environment*: New skills should be learnt in as near as possible 'real' conditions, though of course without the distractions of everyday life. Pilot training flight simulators are an ideal example of a good learning environment. Trainees should be encouraged to experiment in making mistakes, to crash the plane from time to time. As the Americans say: 'People who never make mistakes, never make anything.'

5. *Practice*: The more actively the student is involved, the better. Knowledge should be discussed, difficulties experienced and solved, and skills practised, as part of the learning process.

6. *Involve several senses*: Most of our learning comes through sight; if hearing or touch or other senses can be involved too, the learning process is reinforced.

Training methods

Modern training tries to use as many different techniques as possible.

1. *Classroom training*: Using well designed aids such as overhead projector slides, blackboards, flip charts and so on can be a very effective way of getting trainees to step back from the day-to-day problems of the company. The unfamiliar classroom atmosphere can be both a help and a hindrance in this. To get the most out of this technique, there should be a clear course outline, with written materials, constant trainee involvement in the course, and trainers who not only know the job, but can also communicate well. Attention should be paid to the pacing of the course; trainees should not be asked to sit for long periods listening without participating.

2. *Training films*: The Irish Quality Control Association maintains a library of video training films for use by members. These cover all subjects relevant to quality from audits to standards. The AnCO Open Learning programme called *Total Quality Control* comprises a set of films and booklets designed as an introduction to quality control for small companies.

Training for quality

Training requires the provision of adequate facilities R. &. A. Bailey

3. *Audiovisual aids*: These consist of a set of slides connected to a taped voice-over, the purpose of which is to let the operator follow the training programme at his or her own pace. There are standard programmes available for operations in general use such as electronic component assembly and tool maintenance. Because each trainee goes at his or her own pace, this can be a very efficient way of receiving complex information, especially since the messages of sight and sound can reinforce each other.

4. *Programmed instructions*: The booklets attached to the AnCO course *Total Quality Control* are designed by the programmed learning method. Each topic is divided into small segments or frames of knowledge. The student reads, learns and is tested periodically on his or her understanding of that frame before progressing on to the next section. The student then becomes confident that the knowledge has been grasped before the next segment is tackled. Considerable attention has to be given by the writer of programmed learning courses to the test questions. If they can be answered from the short-term memory without necessarily having been fully understood, the method will fail.

5. *Books and magazines*: The printed media have an important role to play in

Training in robotics *AnCO*

any training programme. Books and magazines are often the easiest and most palatable way of acquiring and transmitting knowledge. They are also invaluable in maintaining current awareness of industry standards. This will become important when the new product liability rules (see Chapter 28) come into force. One defence against liability is likely to be that the product was tested by as advanced a technology as was feasible. Companies will need therefore to keep an eye on recent developments. This will mean the setting up of more formal company libraries, which will act as training resources as well.

6. *Simulations*: There are two kinds of simulations. Mechanical simulations and games involve the use of models and games rules to enable the trainee to experience a particular situation. For instance, what happens in sampling theory can be simulated with sets of marked chips and a bowl. If frequent selections (with replacement) of the chips are made, a sampling distribution can quickly be demonstrated, and how that sampling distribution approximates to the underlying data. A few packs of cards, well shuffled, with varying numbers of aces in them, can be used to demonstrate how sampling plans work. More elaborate and sophisticated statistical training simulations are available. The other

kind of simulation is by role-playing scenarios. By acting out certain roles, with given characteristics, the trainee receives a vivid impression of the human forces underlying a situation. In one example, worker and management representatives on a company safety committee were asked to act out each other's role, with very interesting results.

7. *On-the-job training*: Sometimes dismissively called 'sitting with Nellie', this form of training can be extremely valuable. It is given in the normal work situation, with the machines that are to be part of the working day, and involves virtually no element of unreality. The supervisor should ensure that the training objectives are well understood, and that the matter to be taught has been well prepared. The instructor should be both knowledgeable and enthusiastic about the work. The process is first explained and then demonstrated. The trainee should describe what is to be done, and then either watch another demonstration, or do the task. In either case it is important to reinforce the practical demonstration with the spoken statement.

Training objectives

The purpose of training is to fill the gap between what is wanted and what is being supplied. The gap can be detected in two ways: by the flaring up of a problem, or by ongoing performance analysis. If the level of customer complaints suddenly jumps, there is a problem. The returned products will help the quality assurance staff to identify it. It may be in the raw materials, in which case the quality standards for supplier assessment need to be revised, or perhaps operators and inspectors need further training in spotting this fault. In many cases, since problems rarely have only one cause, both the standards and the training will need adjustment.

Production and quality standards are set to ensure that the production process runs correctly. To achieve these standards, a continuous programme of assessment and training should be implemented. As staff leave and are replaced, as new skills are required, as new products are introduced to the line, as existing staff need to be upgraded and remotivated, training programmes should be devised and effected accordingly.

Training in quality disciplines is a specific part of the quality plan. The objective is to instil two things into the whole body of employees (not just the quality assurance staff). Of course the achievement of quality depends not on the mugging up of a few quality skills, but on the training of all relevant employees in their respective jobs. There are, however, two aspects of the quality approach that deserve special attention.

1. *Quality disciplines*: This is the body of knowledge required to put the quality objectives of the company into practice. Just as a different understanding of finance

is required at different levels of the company, so different quality disciplines will be required in the same way.

2. *Quality motivation*: Knowledge without motivation is useless. It is therefore a prime objective of the quality training programme to demonstrate the desirability of the quality approach.

With these two broad objectives in mind, the quality training plan can be devised. This should include answers to the following questions:
— Who is to be trained?
— What are the precise objectives of the training?
— What sort of training is most suitable: in-house, on-the-job, or publicly available?
— How long will the training period be?
— What will the cost be, including the cost of covering for staff absent on training courses?
— What courses, materials and course leaders are available?

The quality manual should lay down both the standards of education and training required for any post, but also the training techniques to be used to achieve those standards. A typical quality training programme for supervisors would include the following elements:

1. *Supervision*: The customers; how the company is organised; how the various departments relate to one another; the principles of supervision; responsibilities of the supervisor; work planning and assignment; motivation; cost control and budgeting.

2. *Quality policy*: The quality policy of the company; safety and product liability; the quality manual; standards; the quality department; audit, inspection and test systems.

3. *Job technology*: Specifications; materials; processes; products.

4. *Measurement*: Test specifications; metrology; instrumentation; maintenance of accuracy.

5. *Statistics*: Averages; variation; distributions; sampling for attributes and variables; use of sampling plans; control charts; process capability.

6. *Conformance*: Decisions on conformance to standards; deviation procedure; materials review board; disposition of non-conforming products.

7. *Troubleshooting*: Investigation of shop troubles; investigation of field complaints; basic problem solving techniques (80/20 analysis, cause and effect diagrams, histograms, graphs).

8. *Inspection and test data*: Data collection; analysis; feedback.

Without training, the company cannot develop. The training programme should cover the whole staff, from top management to the newest recruit. Each level should be trained in the quality disciplines and the quality motivations appropriate to it. The important point is that planning, and therefore training for quality, involves the whole company.

The IQCA training programme

The Irish Quality Control Association organises a range of courses and training programmes in conjunction with AnCO and various third level colleges. The full set of programmes consists of:
— forty-eight regional branch evening meetings/lectures per year
— an average of twenty courses run each year in various locations throughout the country
— a Certificate in Quality Control course run in many third level colleges
— a Certificate of Management of Quality Control course also run in third level colleges
— a two year full-time National Diploma in Technology (Quality Control) course run by Sligo Regional Technical College
— a one year postgraduate Diploma in Quality Assurance run by University College, Galway
— a two year postgraduate Diploma in Quality Assurance run by Trinity College, Dublin
— a home study course in Statistical Quality Control.

Follow-up

As in any control system, the training programme must include a feedback function. In other words, an essential part of the process is to check whether the specific production, quality or motivation objectives have been met. Are the desired new standards now being achieved? This check should be done reasonably promptly after the training programme ends, so that any deficiency in the programme can be quickly corrected. The normal production and quality controls should be able to reveal whether the training succeeded or failed to 'stick' in the longer term.

Finally, records should be kept of all the training given to each individual. This will help in the setting up of future sessions, and will provide a store of expertise that may be needed later.

Chapter 15

Quality circles

MANY companies have had the vague idea that there is an untapped reservoir of information and skill in their workforce. Some make weak attempts to tap this reservoir by the use of Suggestion Boxes, etc. The impact of these boxes has usually been feeble at best. A common management attitude to them was revealed in the *New Yorker* cartoon which showed the Suggestions Box placed in the corridor, directly over the waste-paper basket.

Recently there has been a concerted effort to set up systems to use the latent creativity of ordinary workers. They, after all, feel the impact of quality problems first, and are often much closer to the causes of them. The method used is called 'quality circles'. This idea originated in Japan: the first quality circles were set up there in 1962, and by 1980 there were 120,000 registered circles in Japan, with as many as ten million workers involved in them. The idea spread to Europe and America in the late 1960s, and now many companies all over the world have them. By 1986 twenty Irish companies had seventy active quality circles, and many more have experimented with them.

The idea is simple. Each quality circle consists of between five and seven workers, led by their supervisor, who meet regularly and voluntarily to consider day-to-day production and quality problems and suggest solutions to management. They use basic problem identifying and solving techniques (such as those discussed in Chapter 20). The solutions they recommend are then presented to management for adoption. Successful operation of quality circles has numerous benefits. The most straightforward is the financial benefit. Rolls-Royce, not a company known for poor quality, saved £248,000 in the first eighteen months of their quality circles' operation. Other practical areas of benefit are:
— improved quality
— improved safety
— better communications
— improvements to systems and processes
— improved industrial relations
— energy conservation
— reduced absenteeism.

As well as the practical benefits, there are enormous potential psychological benefits. The primary force operating in a successful quality circle is the involvement of the individual worker. The philosophy builds on modern theories of motivation which stress the necessity of job satisfaction by enabling people to show creativity. The underlying assumption is that quality is not achieved by white coated quality control inspectors armed with manuals, but by the supervisor and the workers on the shop floor. If they don't get it right, no one will. The theory states that everyone is capable of making a contribution towards improving efficiency, and that a tremendous amount of specialised knowledge about quality and production problems is available on the shop floor, and that this is one way of tapping that knowledge.

The quality circle idea has been mainly a factory floor phenomenon, but it has applications in other areas too. Quality circles exist in banking, clerical and administrative functions, in canteens, in data processing departments, in sales, secretarial and design areas. Because the idea is to some extent a reversal of the old-fashioned attitude to factory work ('You're not paid to think, you're paid to obey orders'), quality circles have undoubtedly had their greatest impact in factory situations.

Elements of a successful quality circle

1. *Management attitude*: Quality circles are made or broken by the attitude of management. Those schemes that have not worked in Ireland have usually failed either because management was insufficiently supportive, or because they interfered too much. Quality circles can be seen as a threat. They are after all suggesting solutions to problems that management has presumably either not noticed or failed to solve. They move a centre of initiative from managerial level to supervisor level. On the other hand, if they work, they can provide management with a major new resource, in the form of a creative and motivated workforce.

2. *Voluntary participation*: All members of the quality circle should be volunteers. No one should be forced to join. This applies to the members of the circles as well as to the managers and supervisors in whose department they work. The whole ethos is one of voluntary cooperation and creativity. Forced participation will create bad attitudes. In fact experience has shown that if the manager is enthusiastic, as many as 80 per cent of the staff of a department will volunteer. They may not all stay the course: the average turnover rate is between 10 and 20 per cent every year. Voluntary participation is essential to the key objective of the circle idea, which is to develop people, not simply to use them.

3. *Training*: Dr Ishikawa, the father of quality circles in Japan, wrote, 'A ton of enthusiasm is worthless unless backed by an ounce of scientific knowledge.' Training should concentrate first of all on problem solving techniques. The more disciplined the approach to a problem, the more likely it is that a solution will be forthcoming. The solutions also have to be presented to management, and groups may need to be taught how best to do that. The second area of training

is in interpersonal relations. The quality circle has to work as a problem solving team, which may be unfamiliar to its members. Relations with management and fellow workers need to be carefully considered.

4. *Solving problems*: Quality circles are a formal way of providing solutions to production and quality problems. They are not merely an early warning system for problems which management then try to solve. It is central to the idea that the circle sees its suggested solutions in operation.

Exhibit: A quality circle solution
Part of Tony's job as general handyman was to go round the plant once a quarter making sure that all the screws on the various machines were snug and tight. He also had many other things to do, some of which were urgent, and which were allowed to interrupt the screw-tightening schedule. The problem then was to remember which screws had been tightened and which not.

The problem was presented to one of the factory's quality circles. After a brainstorming session, during which area- and paper-based solutions were discarded, an ingenious solution was found. Before he tightened each screw Tony was to dip the screw-driver into a tin of red paint. Tightened screws would then be bright red, and the untightened ones dull.

Quality circle components

A quality circle system introduces four new roles into company organisation. These are: the facilitator; the circle leaders; the circle members; and the specialists. The people in these roles have of course constantly to relate to the normal managerial structure of the company. Usually, the idea for quality circles will have come by someone reading about them, or perhaps having been to a seminar on the subject. The idea is discussed by the board, and a decision made. It is vital that management be committed to the idea, and express that commitment. Stories are told of managers who were enthusiastic, but didn't express strong interest. As a result the people in the department lost interest and the idea faded. One manager in this situation commented, 'Obviously I am in favour of quality circles, otherwise I wouldn't have them in the department.' Unfortunately it is not enough just to let things happen.

1. *The facilitator*: This is the key person in the operation of quality circles in a company. His or her time represents an investment and a commitment by management to the project. The facilitator's role is to promote the quality circle idea in the company. This involves setting up the pilot circles, probably with the help of a consultant; devising and carrying out training programmes; coordinating circle operations; acting as link-person between different circles; liaising with management; maintaining records. The facilitator is the link between the circle and the rest of the organisation. Ideally facilitators should be full-time, so that they can apply themselves totally to the project.

Quality circles

2. The circle leader: Quality circles operate completely within the existing structures of the company, using the same organisational structure, chain of command and authority patterns. The circle leader should therefore be the supervisor, leading hand or foreman of the group. Putting anyone else in the position of circle leader does not work; the circle then starts with the major problem of having to relate to the supervisor. If the supervisor isn't part of the solution, then he or she will undoubtedly be part of the problem. At the very simplest, if the supervisor is the circle leader, there should be no problem in arranging for circle members to attend meetings. One of the objectives of quality circles is to help supervisors to create team consciousness among the people working for them. Studying and solving mutual problems in an atmosphere away from the hurly-burly of the shop floor can generate just that.

3. The circle members: These are the basic resource that the project is intended to develop, the 'gold in the mind', as Rolls-Royce managers put it. Membership of a circle must be voluntary, and need not be permanent. It might make sense for the circle membership to change as different problems are being studied.

Quality depends on the individual operator   Howmedica

Obviously as many people as want to should be given a chance of membership of a circle. In general it is found that quality circles work best if all the members are from the same work area. There are occasions, however, when people from different departments, who perhaps share the same problem, might usefully come together. It may be that the facilitator will see that the combined forces of two circles might come together on a particular problem. There are no rigid rules; like everything else, the best circles have the combined loose/tight properties that enable them to change in the face of circumstances without losing identity. There are also no hard and fast rules about the number of members a circle should have. There need to be few enough to make an effective team, but enough to allow for a flow of creativity without being handicapped by absentees. Between five and seven members is the recommended number.

4. *The specialists*: From time to time the circle will get stuck on technical or specialist problems. Specialists should then be called in to help. It might be an accountant, a salesperson, an engineer — anyone who can supply a special expertise that the group lacks. The specialist must be allowed to act only as a consultant, not as a problem solver. If the specialist solves the problem, the whole reason for the quality circle disappears.

Introducing quality circles

Once a company has decided to introduce quality circles, the first thing is to inform management and trade union officials. Both these groups need to back the plan if it is to succeed. A set of presentations and perspective papers should be prepared so that everyone knows exactly what is being proposed. Line management has to be motivated to take an enthusiastic part in the operation of the circles, particularly since managers could see them as usurping some of their functions. From the union's point of view the scheme may initially be seen as yet another management gimmick. The quality and cost-saving objectives of the scheme have to be made clear, as does the totally voluntary nature of the involvement. It is important that the circle members do not feel they are regarded by the union as management toadies of any sort. Although people or wages or work conditions are usually specifically excluded from the quality circles' brief, it can happen that they are regarded by shop-floor workers as a major channel of communication to management. Naturally there will be pressure in this case to use the channel to air grievances. It is an important part of the facilitator's task to prevent this occurring.

The next step is to appoint a steering committee, with management and union members. The first task of the steering committee will be to appoint a facilitator, and then to work out the schedule of implementation together. The chosen facilitator will almost certainly have to undergo some special training for the new job. Once this is done, a draft manual for the operation of quality circles should be prepared. This will serve as the basis for the development of training programmes for circle leaders and members, and also to a lesser extent for managers on how to relate to quality circles. Another aspect that should be considered at this stage is the awareness of the programme throughout the company.

The final step is the selection of the first circle leaders. It is usual to set up a pilot circle, which works from the draft manual, if necessary suggesting changes to it. The circle leader and the members should be trained, and should start work on real problems as soon as possible. Regular feedback to the steering committee should be initiated. As soon as the pilot circle has completed its first operations, and any lessons have been digested, the programme should aim to set up quality circles as rapidly as possible.

How quality circles work

Quality circles usually meet once or twice a fortnight to discuss and solve problems. The problems can relate to cost or waste reduction, to energy saving, safety, equipment, procedures and work methods, communications or quality. Circles are usually specifically told that pay, personalities, products or employment policy must not be considered.

The first step is to identify problems. These may come from the circle itself, or may be suggested by management or the facilitator. Usually it will be easy to create a list of matters that the circle could consider. The second stage is to choose one of the list of problems to start work on. It is important that this be chosen by the members, not insisted on by management; after all, there is little point in insisting on an all-volunteer circle and then dictating to it.

Once a problem has been identified, it has to be analysed. The problem solving techniques discussed in Chapter 20 describe how to go about this. Part of the analysis stage is that the group, having selected a problem, should start tackling it. They have to take a positive role in respect of their chosen problem. It is no longer enough to say to management — Here is a problem! On the other hand, this is for most workplaces a new demand on the workers, and so mistakes will be made. The better understanding of the complexities of any particular problem is part of the benefit.

Quite often the analysis of the problem will throw up areas the group knows nothing about. A circle of machinists might hit an electrical problem, for instance. A specialist is then called in, and asked to provide the necessary extra knowledge. The natural outcome of all this analysis and information about the chosen problem is a set of solutions. In many cases these will be within the authority of the supervisor to implement, so action can be taken at once. If the solution is more far-reaching, it will have to be presented to management. This is a crucial stage in the cycle. The members of the circle will be proud of their work, no doubt, but also apprehensive that they may have omitted something, or that their solution will be rejected. Management are also being put on the spot. They must consider the solution not only on its merits, but also in the context of the quality circle itself.

No system should be set up without a feedback stage. Circles should be encouraged to monitor the success or otherwise of their suggested solutions. Not all of them will work, and it will be important for future problem solving to consider why.

Quality circle problems

Although quality circles have been a triumphant success in Japan and in companies all over the world, they have also failed in many cases. There are various reasons.

1. *Lack of management support*: Management support and interest must be real and demonstrated. This applies both at the problem finding and analysis stage and at the solution implementing stage. If a recommendation is rejected, it must be done with as much consideration as the circle showed in suggesting it.

2. *Coercion*: Circle leaders and members must have complete freedom to join or withdraw as they choose. After all, the purpose of the scheme is to stimulate and use their creativity. This cannot be done in conditions of pressure.

3. *Over-ambitious projects*: The facilitator should take care that circles do not tackle problems that are too big to be successfully resolved. The important thing is to build a record of success, which can best be done with small manageable projects contained in their work area. The pace of the circle's progress should be carefully controlled. The facilitator should not allow a hunger for results to lead to snatched solutions, but on the other hand excessive study and analysis can paralyse the group.

4. *Elitism*: Non-members are an important constituency; they have to put the proposed solution into effect. If they are alienated from the circle, they will be alienated from the solution proposed by it.

5. *Lack of appropriate training*: If the leader and the members of the circle are thrown at a problem without adequate training and practice in the use of problem solving equipment, they will naturally get lost. This will quickly lead to resentment and the circle will be abandoned. Another important training aspect is to control the type of problem that the circle considers. A quality circle is intended to improve the quality of work, not the working conditions. The group must train itself to concentrate on work-related issues.

6. *Lack of commitment*: Everyone, including top managers, middle managers and unions, must be committed to the scheme. An important part of this commitment

is the willingness of management to supply information to the circles, and to allow time and space for them to operate. They should, for instance, be allocated a special room in which to work on their problems. But it is most important of all that the facilitator be enthusiastic. If the person promoting the scheme regards the whole thing as a chore, the idea has no chance.

7. *Personnel problems*: The operation of the circle must be conducted inside the normal authority structure of the company. This can exacerbate problems that already exist between managers and supervisors, or between supervisors and workers. These conflicts must be resolved before progress can be made.

Conclusion

The quality circle idea has aroused scepticism. Many see it as yet another short-lived management gimmick. However, the basic idea, using the creative energies of the workers, is sound; but it is not a substitute for a quality plan. Indeed the idea can work only if a satisfactory quality plan and the attitudes that such a plan implies are already in operation.

PART III

The quality toolkit

Chapter 16

The quality manual

FOR eight hundred years, stimulated at first by frequent civil wars and then by military rule, Japanese craftsmen made the finest swords in the world. The skill died only after the samurai were forbidden to carry swords in 1876. Each of these lethal weapons was made of two grades of steel welded to an iron handle. The steel was folded in on itself, fired, welded and hammered out to its original length. Between each new welding, the blade was coated with clay. This process was repeated up to twenty times until there were many thousands of layers of steel. Sometimes three or four sheets were welded and folded together, finally producing more than five million layers. The blade was shaped, honed, polished and sharpened for more than fifty days. The finished products were immensely strong. A training film shown to Japanese troops in the Second World War showed an expert swordsman cutting through the barrel of a machine gun. Quality control rested on the master craftsman's jealousy of his own reputation, and also on professional sword-testers.

The swords were made by a craftsman who had spent years learning the secrets of metalwork, of polishing and honing the blade, balancing the sword and other skills. So great was the personal and technical identification that craftsmen often took the name of their master for their own. Everything was learnt by example: nothing was written down. All the traditional crafts of the world, from saddle-making to porcelain manufacture, were passed down in this way from teacher to pupil. Certain areas, where an especially good teacher lived, became famous for particular types of craft product. Cheese, chairs, clocks, knives, carpets, lace and glassware all became area specialities in this way.

Not even in Japan is it nowadays practical to undertake the old craft training system. But firms still have to produce goods to the highest accuracy and quality specifications. The old craft system of person to person learning takes too long and is too inefficient. So we establish a set of written procedures which everyone is asked to follow, and which cover every function that might affect product or service quality. A collection of these standard operating procedures is called a *quality manual*.

What is the quality manual for?

The quality manual is a method of achieving the highest standards of quality production without laborious and lengthy personal tuition. Its words and diagrams replace the vagaries of personal training — 'sitting with Nellie' — with clear how-to rules and standards. Personal tuition is not replaced by the manual, however, but is made doubly effective by having a firm base.

A quality manual therefore is:
1. a bank of knowledge about production practices
2. a summary of standards, for internal and sales use
3. a means of indicating the personal and departmental relationships in the firm
4. a base for the activities of the quality control staff and for the quality audit and review procedures
5. a statement of policy relating to quality
6. a direction-finder for the training programme.

Obviously, the more complex and extensive the activities of the firm are, the greater the need for a quality manual. As the quality programme discussed in Chapter 1 gets into its stride, the project team will produce memos, notices and recommendations about quality matters. If the memos and notices are produced independently of each other, sooner or later one will contradict another. For instance, a suggestion for improving postal department service (by increasing the number of internal letter deliveries at the expense of parcels) may make it impossible for the customer complaints people, who usually receive the complaints by parcel, to achieve the recommended standard of a same-day turnround of response.

The practical origin of the quality manual is the need to ensure that each recommendation fits into an overall quality plan. The next problem is how to turn a haphazard collection of information about practices, standards and responsibilities into a coordinated manual.

Before starting

Before going into the preparation of the manual the team needs to consider various general questions.

1. *Who is going to use this manual, where, and for what purpose?* Clearly a manual intended mainly for the quality assurance team can be written and presented in more technical detail than one intended for general management and shop-floor use. The manual is primarily a focus of communication. As such, it must be carefully prepared to suit its audience.

2. *What level of detail should the manual contain? Should there be a shortened version for visitors?* One major use of a quality manual is to reassure key customers that the quality standards specified will be maintained. But to be of any use inside the company, the manual must contain significant, and perhaps confidential, details

of the company's working practices. Many companies get over this difficulty by maintaining two manuals, one for internal use and one (abbreviated) for customers. Linked with this question is the problem of how distribution inside the company should be controlled. Should there be a statement of who has a copy, and who is to have access to the copies? The fewer the copies, the easier they will be to amend and update, and the more confidential the firm's quality practices will be. On the other hand, the quality plan will work only if it is a common possession of all the company's employees, at least in some form.

3. *How will suitable editorial standards of writing and communication be maintained?* Technical English is notoriously difficult to write clearly. Anyone who has tried to decode operating instructions that were badly translated from the original Italian or German will know this. It is not merely a matter of using simple language. Language that can be understood by a nine-year-old is dangerous if there is one ambiguous sentence which enables a safety valve to be installed upside down. A particular problem with a quality manual is that many potential users will assume that it is not worth reading. Studies have shown that many people prefer to ask others for information rather than look it up in a manual; some will assume that they know what's in it already; others will prefer to muddle along as they always have. A key task of the manual editor therefore is to make the contents palatable and easily accessible to its readers. It might be desirable to hire an outside consultant to be sure of getting this aspect right.

4. *How should the manual be presented physically?* Typeset material is easier to read than typewritten text, and is as much as 40 per cent more memorable. It is also much less bulky. On the other hand, it is expensive and less flexible. The physical layout of the text should also be given careful consideration. One simple rule of thumb used by typographers, for instance, is to ensure that the space between the lines is greater than the average space between the words. Most word processors allow you to produce 'ragged-right' text by switching off the right hand justification. Surprisingly this makes text easier to read, because equal spacing between words is a critical factor in legibility. Clearly laid out tables and diagrams can also significantly improve ease of reading. If the pages are to be handled in oily or dirty environments, perhaps they should be individually laminated? The system of binding will depend very much on how frequently the team expects to update the material. The most common option is a ring-binder system.

5. *What system should be established for updating and maintaining the manual?* No system stays still. To maintain usefulness and credibility, the manual must keep up with changes in the company's production patterns. It is important to envisage how this will be done right from the start. How are the new ideas to be assessed, written up, reviewed and formally incorporated as company policy? How will

QA staff using the quality manual during testing procedures   R. & A. Bailey

the updates be integrated into the contents, the index and the main text? How will the new material be distributed? How will the quality team make sure that every manual has been updated, and that the new systems are in operation?

6. *How can the quality team ensure that the systems described in the manual are operative and supported by the highest level of management?* As soon as the recommendations in the manual become detached from real life, the quality system will collapse. If the supervisors find themselves saying, 'Never mind what it says in the book of rules, lad, just you do what I tell you,' that is a major danger

signal. Either the manual doesn't reflect reality, or the supervisors haven't been sufficiently convinced of the necessity for a quality plan. Typically this occurs when salespeople or senior management set targets that make quality production impossible. The quality team can achieve results only if it ensures that the quality manual continues to reflect reality in the boardroom and on the shop floor.

7. *Is the manual to be a statement of existing policy, or a persuasive document concerned with the introduction of a new policy?* A manual that simply formalises existing rules is quite a different document to one that is introducing new standards. In the first case, the presentation should concentrate on ease of access, rather than persuasive presentation. The style of introduction of the manual should be different, too. If the manual is the spearhead of a new policy, careful training/induction programmes would have to be devised, at every level of the company, to ease the new systems into action.

In general the style of the manual must continually be related to its potential use. In some circumstances an elaborate multi-volume reference work, designed for occasional use by the quality control staff, may be appropriate. In other cases a more simple document for the staff to use daily would be more effective. Invariably the actual manual will have some of the flavour of both approaches.

It may be desirable to split the manual into sections. Some of these, such as those containing general policy statements and practices and the more permanent features of production practice, would be widely circulated. Others, such as those containing quality assurance procedures and the more confidential aspects of quality practice, might be restricted on a 'need to know' basis.

Exhibit: Compiling a quality manual at Irish Fher Laboratories
The idea of compiling a quality manual at Irish Fher first came about in late 1982. Given that there was a quality system of a relatively high standard already in operation, the main advantages were seen as follows:
– defining individual responsibilities in a clearer and more precise way;
– cataloguing some of the existing procedures and ways of doing things;
– motivating individuals;
– putting across the quality message;
– improving and strengthening the whole system;
– acting as a standard against which procedures would be audited.

Sources of information
The most useful source was the existing written procedures. After that, a quality manual from another chemical plant was helpful, although we were allowed only limited access to it because of plant confidentiality rules. Various IQCA conferences and printed sources were also useful.

Problems encountered
From the start the magnitude of the task was underestimated. With hindsight, the necessity of defining the sections and listing the headings at an early stage cannot be over-emphasised. Only this way can the group gauge the technical and secretarial resources necessary to complete the job. The whole

process of assembling the manual took approximately twelve months, and appeared to be a never-ending, insurmountable task.

It was decided at the outset that the sections of the manual should be stored in a word processing system. This decision was made in order to facilitate speedy and efficient updating of any section. Unfortunately the lack of a firm contents plan led to difficulties with storage space and ultimately with access to certain sections.

Getting agreement between various sections as to responsibilities was a slow process.

The question of whether to document every procedure in the manual or merely to make reference to it in the manual was debated. Eventually the latter option was generally chosen, but ideally a master file with all operating procedures should be set up.

Conclusion

Compiling a manual is a long and arduous task. However, the benefits to the organisation are enormous. In particular, the manual provided unexpected benefits in:

— defining and making explicit inter-departmental responsibilities. Departmental managers were asked to sign the relevant sections in the manual to ensure commitment

— providing an essential benchmark for the start of internal auditing

— focusing attention on the instrument calibration system, which was vastly improved as a result.

The lessons learnt were:

— the company should allow 12-18 months for the exercise

— the list of sections should be defined as early as possible

— the widest possible participation from all levels in the company should be encouraged. Irish Fher still has some work to do to increase understanding and ownership of the manual

— problems have been experienced in revising and updating the manual. It is suggested that the manager of the section should be responsible for this rather than the quality control manager.

Source: Dr Conor O'Brien, Chemicals Department Manager, Irish Fher Laboratories, Co. Cork

However it is done, the quality manual must first of all clearly define the quality system. Secondly, it must demonstrate clearly the systems aspects of structure, content, communication and control, so that everyone knows what is expected of them. And finally it must itself be a quality product — nothing carries less conviction than a tatty, ill-written, ill-designed manual preaching quality in all things.

Structure and content of the manual

In practice, the compilation of a quality manual will work in two directions. From the bottom up, because the company will already have various quality rules and practices, and from the top down, in that the quality team must impose an overall structure on what's already there.

At an early stage, the quality team will need to draft a contents list for the manual, which will enable them to see how existing house rules fit into that structure. This contents list should cover every aspect of the proposed manual. The list is likely to change slightly as the work progresses, but the basic frame should be adhered to if possible.

The quality manual

Different organisations will require different contents structures. It is not possible to recommend one structure that can cover such diverse enterprises as a large general food store like Quinnsworth, an architects' practice, and a contract manufacturing company. On the other hand, every good quality manual has some elements in common with others.

Every section of the manual must describe:
1. the relationships and communications between departments necessary to achieve quality standards;
2. the policies and standards that govern the work; and
3. how the output is to be reviewed and evaluated.

Obviously the exact content will vary from company to company and from department to department. The methods used in evaluating the 'output' of the purchasing department will be very different from those used on the shop floor. In the production of skin creams and medicines, for instance, the most exacting chemical tests will be essential; for matches and tissues not much more than a sample count and use-test need be done.

This basic structure should be followed for every department, so that everyone in the company is clear how the quality plan affects them. In addition, a good manual will also contain certain general statements about the company and about the manual itself. These typically include:
1. a statement of policy from the chief executive;
2. a statement describing the degree of confidentiality of the manual;
3. contents, index and a statement of 'how to use this manual';
4. a brief historical description of the company and its present situation;
5. a section describing how the manual is controlled, describing how copies are to be circulated and how new procedures and amendments are added, and other details of documentation procedures;
6. a glossary section describing terms used in the manual.

Exhibit: Typical quality manual contents
1. Scope
2. General principles
3. Definitions
4. Management objectives
5. Programming and planning
   5.1 General
   5.2 Programme
   5.3 Planning
6. Principles of control
7. Design/specification control
   7.1 General
   7.2 Detail design
   7.3 Manufacturing/production specifications

    7.4 Safety
    7.5 Reliability
    7.6 Development
    7.7 Design/specification review procedures
 8. Purchasing control
    8.1 General
    8.2 Purchasing data
    8.3 Supplier appraisal
    8.4 Receiving (incoming) inspection
 9. Manufacturing/production control
    9.1 General
    9.2 Process/material control
    9.3 Process capability
    9.4 Production tooling, gauging and test equipment
    9.5 Inspection
    9.6 Sampling procedures
10. Marketing/servicing
    10.1 General
    10.2 Marketing
    10.3 Servicing
11. Documentation
    11.1 General
    11.2 Work instructions
    11.3 Modification control
    11.4 Records
    11.5 Quality manuals
12. Defect/failure analysis and remedial action
    12.1 General
    12.2 Defects
    12.3 Failures
    12.4 Analysis and remedial action
13. Control of non-conforming material and components
    13.1 General
    13.2 Segregation procedures
    13.3 Concessionary procedures
14. Review and evaluation procedures

Conclusion

The production of a good quality manual can take months. It is expensive in executive and worker time. A bad manual can be done much more quickly and cheaply, but will soon be found collecting dust on the tops of cupboards. However sooner or later there will come the stage in the development of the manual when senior management want the thing finished '*now!*'

To prevent this from happening too early, and to reassure management generally, it is a good tactic to provide immediate benefits from the work on the quality manual. This is a common tactic, akin to the bell-wether project idea discussed in Chapter 2. On one occasion, a management consultant was asked to undertake a major reorganisation of a large company. To do the job properly would take some months and cost a great deal of money. He therefore looked round in the early weeks for a suggestion that would quickly save the company money. Luckily he was able to change the stationery ordering system, which had fallen into slack ways. The money saved by his suggestion easily saved his fee and more. He was then able to take the proper time to complete the main assignment.

The creation of the quality manual marks a major stage in the development of quality consciousness in the company. Thereafter the stress is on maintaining the systems in the manual, and ensuring that the manual continually reflects the real objectives of the company. There are two ways to ensure that the precepts of the quality manual are kept: the first and most important is the constant minute attention of management to quality; the second is the quality audit, which is the subject of the next chapter.

Chapter 17

Quality auditing

THE difficulty with good resolutions is — keeping them. After New Year most of us start wavering within a week; and by mid January we have thought of several good reasons why giving up smoking might be positively injurious to health. After all, as we might say to ourselves, there's a difference between judicious persistence and plain obstinacy, or, as the fat lady said, 'Dieting's dieting, and starvation's starvation.'

The same kind of problem afflicts companies that have resolved to adopt quality assurance plans. For the first few weeks all goes well. The factory and the offices are clean, the charts are carefully filled in, the checks are meticulously done. If management have cleverly chosen to adopt the scheme during a relatively slack period, the introduction will be so much easier.

The first problem comes when business starts to pick up again. That's when the conflict between short-term urgency of production targets and long-term quality standards first begins to show. But if management don't give in too readily to the first temptation, all that might happen is a temporary slight loosening of standards. And that gives rise to the next problem.

Once standards have been eased, it becomes much more difficult to maintain them solidly thereafter. As time goes on, the original enthusiasm for the quality plan becomes a memory, and the quality plan itself, from being an interesting innovation, becomes just another part of the system. Gradually little infringements arise. Somebody gets into the habit of bringing a chocolate bar into a no-food area; cleaning routines are slightly altered; bacteriological checks are done once a fortnight instead of once a week; in holiday periods, people are left on the line too long, so their concentration slips; records are no longer kept properly.

To stop this kind of slippage, management must arrange for the plant to have a quality audit.

The idea of quality audits originated in America, when very large corporations, particularly NASA and defence and food buyers, needed to be sure that their quality standards and policies were being kept. They are now an integral part of most quality management schemes. They are also insisted on by various quality management standards, such as Irish Standard 300: 1984.

Auditing has been defined as 'a systematic investigation or appraisal of procedures or operations for the purpose of determining conformity with prescribed procedures'. This definition was originally written to refer to financial auditing, but it describes the purpose of quality audits just as well. It is the purpose of quality auditing to present to management a true and fair view of how their quality plans work in practice. As for financial auditing, a very useful spin-off of the auditing process lies in the ability of the auditor to spot opportunities for improvements in systems and procedures.

Exhibit: The Irish Standard definition of quality audits
The Irish Standard that describes quality audits (IS 303: 1984) defines a general quality audit as 'a methodical investigation of the Quality situation in relation to a product, process or organisation carried out in co-operation with involved parties but independently of them in order to verify how that Quality situation matches up to the specified requirements and how adequate those requirements are. The "Quality situation" can refer to any or all of the following
(a) Products in the broad sense (i.e. materials, e.g. the finished products from a manufacturing process, services, items of equipment) including everything which is necessary for their use (including documentation), treatment, and maintenance,
(b) Processes,
These cover the progressive stages and methods of operation involved in the creation of all or part of a product or service,
(c) Organisations/Systems.
By this is meant the terms, conditions and specific or general rules on which all or some of the methods employed in manufacturing a product or providing a service, are based, e.g. a system or structure of Quality Assurance, a procedure or a specific instruction.'

Types of audit

Audits can be distinguished by the type or subject of the audited system or by the type of auditor. By system, an auditor might be asked to do a system audit, a supplier audit, a hygiene audit, a Good Manufacturing Practices audit, a product audit or a process audit. Each of these differs slightly in focus and intent. From the point of view of the auditor, there are three types.

1. *Internal audit*: Initiated by the company on its own systems, this type of audit is carried out by an employee or a consultant hired by the company, perhaps on a particular aspect of the quality system that has given cause for concern.

2. *External audit*: This is an audit carried out by a company to evaluate its suppliers, agents or licensees. Its scope is very often specified in the original contract between the company and the supplier.

3. *Extrinsic audit*: Typically this kind of audit is carried out by an independent body such as the Irish Quality Control Association, with a view to assessing how far the company complies with a pre-set standard. The specific requirements

for the Irish Quality Control Association's Quality Mark scheme will be discussed later in the chapter.

The depth of the audit

Before the audit can start, the company must decide what it wants to audit, and in what depth. It might only be necessary to check that the systems of the company are in conformity to a particular standard, in which case the audit will not examine in detail whether the system as laid down in the quality manual is actually being complied with. In this case the auditor might simply make a few random checks on parts of the system to ensure that the quality manual is in general use.

Generally, management are more likely to be interested in checking whether the systems laid down in the quality manual are in fact being followed. This is called a 'compliance audit'. The auditor will concentrate on detailed checks on how the system operates on the shop floor, without commenting on the adequacy or otherwise of the system.

Usually audits will contain elements of the system and the compliance approach.

Another aspect of audit depth is the amount of detail in the investigation. A full audit (often called a 'womb to tomb' audit) covers every aspect of a quality plan in detail. Typically the auditor will follow the progress of the raw materials through the factory, from delivery and storage to warehousing, despatch and on as far as the customer complaints system. Another approach is to follow the product backwards through the system from the receipt of an order to the delivery of the raw materials.

The partial audit covers only certain aspects of the company's operation (those relating to a particular contract, for instance). The follow-up audit is performed to check that the findings of a previous audit have been followed through, and that no slippage has occurred in previously satisfactory areas of operation.

An 'informal audit', like a verbal agreement, is not worth the paper it is written on. The whole purpose of the quality plan and thence of the quality audit is to make sure that the right systems to produce a quality product are in place. This cannot be done otherwise than systematically.

The audit plan

Before the actual audit, the company and the auditor must plan the procedures to be adopted. There are normally five preliminary steps before an audit.

1. *The audit schedule*: This is the description of the series of audits that will be carried out and their frequency and objectives. For instance, the company might like to have a system audit a few weeks after the installation of a quality plan. After that, compliance audits, either department by department or product by

Quality auditing

product, should be planned as a schedule covering the following months. The audit schedule should be laid down at least in outline in the quality manual.

2. *Notify the auditee*: An audit is not a snap check; staff in the department or plant that is to be audited should be notified in advance. This will give them a chance to prepare the relevant information and to allocate staff to conduct the audit team round the plant if necessary. They will also have time to 'tidy up', but this is unlikely to affect the success of the audit, for significant non-compliance with systems can hardly be covered up in a few days.

3. *Obtain and review documentation*: An audit is a review of how far the plant is carrying out prescribed procedures. The audit team must review beforehand all standards, specifications, quality manuals, procedures and instructions, previous audit and inspection reports and the Corrective Action Request file.

4. *Develop the checklist*: Using the information collected and reviewed above, the next step is to create the audit checklist. This is a detailed series of questions covering every aspect of the audit scope. The audit team's job is to see that these questions are answered in the course of the audit. The checklist system ensures that nothing is forgotten in the course of work in the plant. Another advantage is that the detailed work of preparing the checklist — which is often done with

A quality auditor at work   *Fitzwilliam Quality Assurance*

145

the plant's quality manager — gives the audit team some insight into the operation of the plant before they actually start.

5. *Agree the audit programme*: This involves deciding who will conduct the audit, perhaps agreeing the checklist, and discussing with the auditee the actual conduct of the audit. The time of starting and the audit programme should be cleared with the auditee in advance.

The entry interview

Before the actual audit begins, the audit team will meet the various people at the plant who will be involved, to explain the purpose and conduct of the audit. Other elements in the entry interview are:
1. introduce audit team
2. introduce observers, if any
3. discover break and meal times in the plant and times when sections are not in operation
4. check that the quality manual used for the checklist hasn't been superseded
5. if necessary arrange for the audit team to be allocated a meeting room
6. agree on escorts for information gathering stage
7. fix time for final 'exit' interview.

Information gathering

The auditors have to limit their information gathering. Time alone will not allow an exhaustive analysis of everything. The auditor will therefore concentrate on key elements in the process, particularly those relating to product liability/safety and hygiene.

Generally, the auditor will consider four factors: materials, people, equipment and methods. Between them these four control final product quality. In every area the auditor will ask such questions as:

— *Materials*: Are they the correct type? Are they in good condition and suitably handled? Are they all present? Are they identified and traceable?

— *People*: Are they trained and competent? Are they clear on what they should do? Are they sufficiently motivated? Do they possess the physical attributes for the job — eyesight, steady hands, etc.?

— *Equipment*: Is it the correct type? Is it in good condition, well maintained and cleaned? Is it being correctly used? Is it capable of doing the job? Is it in the right place? Is it in the right environment?

— *Methods*: Are they sufficient and clear? Are they up to date? Are the manuals and specifications legible and understandable? Are they distributed to those who need them?

The specific route and style of the audit will be dictated by the planning, in particular as to what kind of audit is required, and the checklist that has been created in advance. But as well as completing the checklist, auditors will be looking for less specific information. For instance the tour of the plant might reveal a task that is not documented in the quality manual, or that employee attitudes are not very positive, or that the general state of plant housekeeping is low.

Information is gathered by looking and by asking questions. This should be done by starting questions with 'how' and 'what', so as to tease a full answer out of the informant. Very little information is gathered by questions that can be answered 'Yes' or 'No.' When someone has described what is done, the follow up, if there is any doubt, should always be 'Show me.'

The quality audit is specifically concerned with how the plant's quality systems work. A large part of the information gathering exercise will therefore be concerned with records, information flow, performance monitoring and checking procedures. The auditor will be especially concerned to assess the degree of discipline with which these things are done.

The exit interview

At the end of the information gathering period, the audit team and the auditee's representative meet to discuss the findings. The purpose of this meeting is to discuss the results of the audit with the management team. The audit team will usually describe their general impressions of the plant and its systems, and talk about any aspects that strongly impressed them, whether for good or ill.

This is in effect a preliminary verbal report of their findings, and can go into considerable detail. At this point management can express their views. The specific judgements of the audit team should be quite clear. For instance, if the audit was to establish whether the plant complied with some standard, the auditor should say at the exit interview whether the plant actually does so or not. Next, the auditor will describe the form and layout of Corrective Action Requests. There will generally be some discussion of these requests, and at the end of the meeting management representatives may be asked to sign the Corrective Action Requests, thereby accepting the problems.

Reporting findings

For management the most important aspect of the audit is the report. This is genuine feedback information, which can be used to improve or tighten up the quality system. A typical audit report will be divided into four parts:

— *Executive summary*: A short over-view of the whole report, highlighting the major findings. The summary should be related to top management's priorities and should be separate from the main report.

— *Corrective Action Requests*: Specific matters requiring corrective action, usually laid out in a standard form, copies of which should be kept by the audit team for use on later audits.

— *Observations*: Brief statements of observations and judgements of the areas audited, both complimentary and otherwise.
— *Audit report details*: A list of audited items with findings and recommendations, normally based on the checklist.

Follow up

An audit should not be regarded as an isolated event. In fact the audit schedule is an integral part of the quality plan. At the end of the audit, the auditor should discuss how the Corrective Action Requests are to be handled, and how the various other general matters arising out of the audit are to be rectified. In some cases it may be necessary to organise a re-audit of the deficient areas.

The Quality Mark scheme

A quality conscious company has to be particularly sensitive about the quality of bought-in materials. After all, things made on the premises can be controlled to be 'right first time' — items made by someone else can't be. As a result, some companies commission audits of their major suppliers as part of the vendor assessment routine. Unfortunately this is expensive and time consuming.

In 1982 the Irish Quality Control Association introduced the Quality Mark, scheme. This is awarded only after extensive and regular auditing organised by the IQCA. One of the key objectives of the Quality Mark scheme is to provide reassurance to purchasxers. Once as company has been awarded the Quality Mark, public and industrial customers can be sure of conformity to requirements. Since the Quality Mark scheme is based on BS 5750/IS 300, companies with the Quality Mark receive a proportional exemption when applying for accreditation under these Standards.

The Quality Mark is not easily acquired. The granting of the mark is based on a twofold assessment of the applicant company's quality control system. There are also follow-up audits each year after the original grant. The first step to gaining the Quality Mark is to complete a questionnaire about the company's quality system. The thirty-one questions ask, for instance, how quality standards are established, who is in charge of quality, what statistical and other quality inspection methods are in force, what vendor assessment is done, how quality costs are measured, and what quality training is given. If the answers to the questions reveal that the company takes quality seriously, an audit of the factory is organised by the IQCA, using independent consultants. The Quality Mark relates only to a single plant; if a company has several plants, each one must be separately accepted for the mark.

The emphasis of the audit varies with the type of product, but in each case the following headings are used. Each has a maximum number of points attached to it, and the company must achieve at least 80 per cent of the possible to gain the award. Even if the company gains more than 80 per cent, it may still fail if the inspection reveals any inadequacy in a crucial area of safety or hygiene, or if the deficiency affects the quality of the finished product.

Exhibit: Maximum marks by category awarded for the Quality Mark scheme

Category	Possible points
1. Quality planning	140
2. Incoming material	150
3. Manufacturing control	240
4. Records	50
5. Environmental control	100
6. Training	150
7. Customer service	50
8. Management of product quality	150

Source: IQCA

After the audit the auditor submits a report to the Approvals Board of the IQCA. This Board consists of the chief executive of the IQCA, as Chairman, and representatives of AnCO, the CII, the ICTU, the IIRS, and the Kilkenny Design Workshops. They consider the summary report produced by the audit inspector, without being told the name of the applicant company.

If the Approvals Board decides to award the Quality Mark to the company, it may then use the standard Quality Mark symbol on all packaging and in advertising. This proclaims to customers that the company's commitment to quality has been independently assessed and verified.

Chapter 18

Metrology

THE technique of mass production, which requires precisely engineered parts to be interchangeable with each other, is the heart of all modern economies. In theory any mass-produced product is identical to those of the same design. One can dismantle half a dozen such products, jumble the pieces, and reassemble them. The idea of mass production actually originated in France, but developed most quickly in the tradition-free United States. The key to mass production is the ability to manufacture, and therefore to measure, increasingly exact quantities. This need has given rise to metrology, the science of measurement.

In 1797 the American government was threatened by war with France, and was conscious of an acute arms shortage — 40,000 muskets were needed. The two national armouries had been able to produce only 1,000 in three years. Each gun was put together by a skilled workman forming and fitting a unique weapon. If a part broke, its replacement had to be specially constructed. The great inventor Eli Whitney responded to the challenge, and designed a machine whereby unskilled workmen could stamp out interchangeable parts that could then be assembled. He compared the pieces to the products of a printing press. All the books in an edition are identical because they are an assembly of identically reproduced sheets. At a famous demonstration in 1801 Thomas Jefferson and members of the American Cabinet picked sets of gun parts at random and assembled their own muskets. As a result Eli Whitney was commissioned to supply 10,000 muskets in two years, or ten times what the national armouries had been able to make in three years.

Engineers soon realised that 'the American system', as it was called, would need a degree of precision in manufacturing that hadn't been required before. If each bolt is made only to fit its own nut, as it was until then, you don't have to worry about uniformity or precision.

Precision

In the early nineteenth century, engineers were capable of working to 1/32 of an inch, but they soon developed precision instruments to work regularly to a tolerance of 1/1000 of an inch, and in 1856 a machine was demonstrated that could measure 1/1000000 of an inch.

Increased precision in engineering stimulated designs and inventions for clocks, scientific instruments and navigation aids. These in turn fostered a demand for more and more precise measuring instruments, and greater precision of standard measures. After all, if the specification calls for a tolerance of no more than 1/1000 of an inch, something more than a simple ruler is needed to check the accuracy of production. As scientific measurement became increasingly precise, scientists became aware that the definition of exact length is not simple. During the Middle Ages, for instance, it was quite sufficient to specify that the standard length for cloth was related, for instance, to the iron bar kept in the city mayor's house. When it was eventually realised that metal expands when hot, it became necessary to regulate the temperature of the bar.

The best way to do this was to lay the bar on rollers; this, however, allows the bar to sag between the rollers, thus very fractionally reducing the distance between the ends. The Victorian imperial yard standard sagged in this way, so that it is now some 200 millionths of an inch shorter than it was when it was created in 1878. British consumers have been getting that much less cloth for their yard since then.

Measurement units

As soon as any kind of society evolves, it needs common measurements and scales. Builders and architects need to be able to specify to quarrymen what size of blocks they need, common weights and measures for bread and beer need to be specified to prevent traders exploiting their customers, and a measure of area is needed to enable land to be bought and sold. Every community has the same needs, and meets them in a variety of ways. The first and simplest standard measure is that of length. In one area in the Middle Ages they lined all the men up on a particular Sunday after Mass and measured their right feet. The average length would then serve as the community's standard measurement for the foot for the rest of the year.

The basic English measures of foot and yard were laid down early on in the Middle Ages, though the foot was generally about 13.2 inches, and the yard either 39.6 inches or 35.96 inches depending on which part of the country you lived in. The longer yard was banned by statute in 1439. In Ireland both the acre and the mile were different to their English counterparts, no doubt causing considerable confusion to the colonists.

The modern *Système Internationale* set of measurements — the metric system — is derived from the system first proposed in 1795 in France. Measurements of mass, length and capacity were interlinked, and the whole system based on the standard metre which was defined as one ten-millionth of the line from the North Pole to the Equator through Paris. Over the years much work has been done to refine this somewhat vague definition. It is now defined as being equal to 1650763.73 wavelengths of the orange-red line in the spectrum of the krypton-86 atom under certain conditions. Napoleon enforced the metric system first in France and then in the countries he conquered. But it wasn't until the 1970s that many other countries such as Ireland and Britain officially adopted the metric standard, and the United States still hasn't.

The science of measurement

The science of measurement — metrology — starts with an idea of something to be measured. This is referred to as the quantity. The next step is to create an instrument to measure that quantity. The instrument has two functions:
— it must enable the operator to compare the values of the quantity (length, weight, speed, etc.) with an objective scale;
— it must enable the operator to reproduce that value, so that the same result will be achieved at a different time from the same quantity. A pair of scales must not only be capable of comparing the weights of two loaves, but also of registering the same weights every day, otherwise today's kilo is different to tomorrow's.

The more subtle the measurement required, the more elaborate the instrument. As instruments of measurement become more elaborate, how they themselves work becomes a factor in the act of measurement. It then becomes increasingly important to ensure that the instruments are giving a true reading.

Measurement can be by various means. These are four common ones:

1. *Direct method*: The value of the quantity is obtained directly, without any special calculations, e.g. measurement of length with a ruler.

2. *Indirect method*: Measurement of a quantity that is known to exist in a direct relationship to the desired quantity, e.g. the measurement of temperature by the movement of mercury in a thermometer tube.

3. *Comparison method*: Measurement of a quantity by comparison to a known value of the same quantity, such as the measurement of a volume of liquid in a measuring jug.

4. *Interpolation/extrapolation method*: Discovery of a value by inference from two or more known values.

Once the instrument has been invented that will give a regular and reproducible reading of a particular quantity, the units of measurement are devised. Ideally the unit should be part of a system of units such as the SI units, which derive all measurements from seven basic units.

Units to measure acceleration, area, capacitance, density, energy, force, inductance, stress, velocity, viscosity and many other quantities can be derived from these seven. For instance, velocity (speed) is derived from distance divided by time. In certain cases, however, such as in the measurement of hardness, the unit has to be derived from an arbitrary set of standards.

Metrology

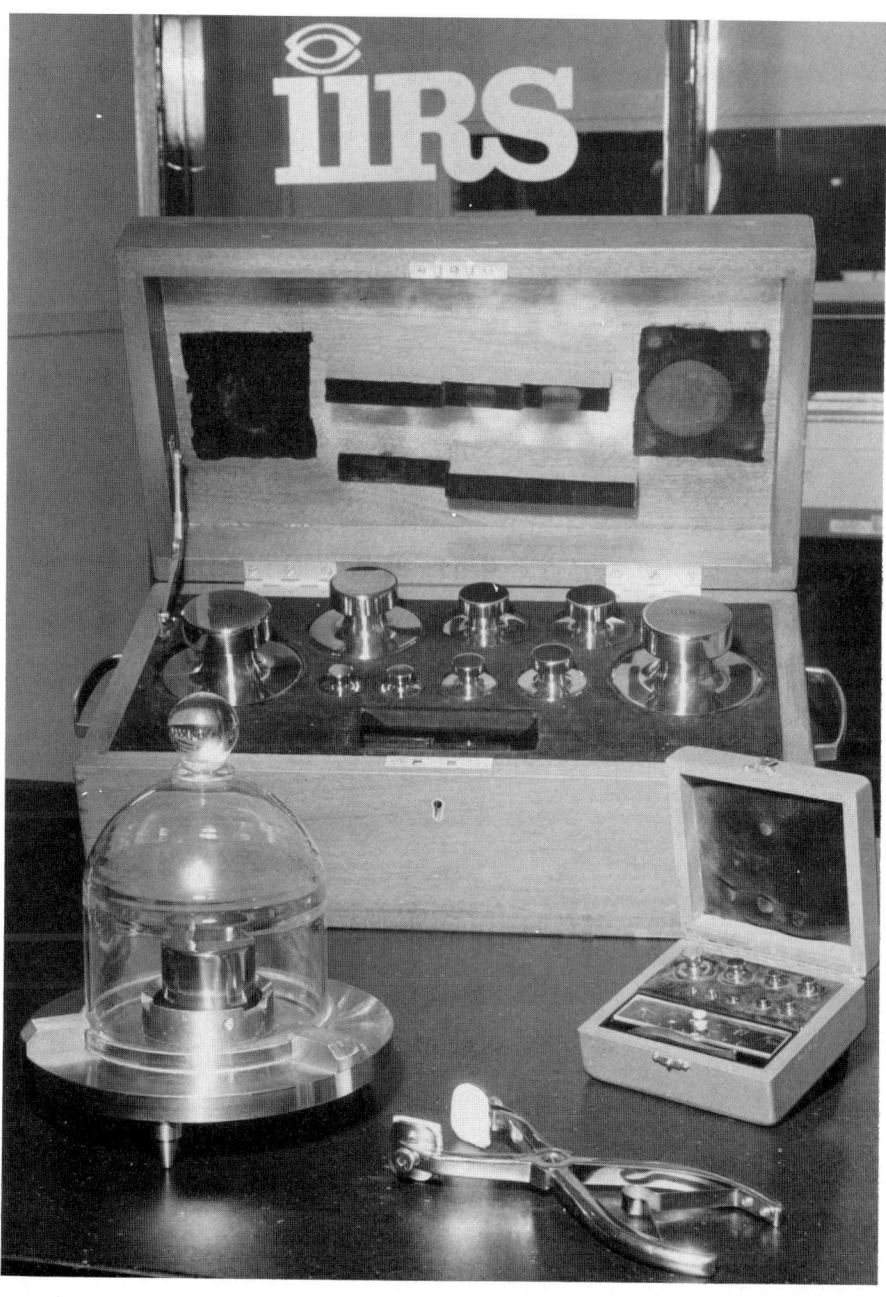

Standardised weights for reference   *IIRS*

Exhibit: The basic units of the SI system

Unit	Measure
metre	length
kilogram	mass (weight)
second	time
ampere	electric current
kelvin	temperature
candela	luminous intensity (light)
mole	amount of substance

Whatever the system of units, the instrument has to be checkable. Either directly or indirectly its results must be able to be compared with the fundamental standard of measurement. This is called calibration. An essential part of the quality plan is the calibration of instruments against a reference standard. A typical reference standard for length is a box of slip gauges, which are slips of metal very accurately sized to particular lengths. These can be calibrated themselves by the IIRS.

Exhibit: Extract from an equipment calibration manual

Nypro Ltd is the largest custom moulder in Ireland. The company is based in Bray, Co. Dublin. In this extract from their equipment calibration manual, the slip gauges are being used to check the accuracy of sets of calipers.

Instruments: Calipers (all types)

The procedure is as follows:

1. Ensure that the faces of the calipers are clean.
2. Visually examine the calipers' faces for corrosion, chipping, wear or any other damage.
3. Check that there is no looseness in the slider. If there is looseness in the slider, it may be adjusted by tightening the two screws on top of the slider unit.
4. Bring the faces of the calipers together and visually check that there is no bowing of the faces.
5. Check that the calipers read zero and mark this on the calibration sheet.
6. Take two measurements using slip gauges. The first measurement should be between 2.01 and 2.49 mm for metric calipers and less than 0.5 inches for imperial calipers. The second reading should be greater than 50 mm or 2 inches as applicable.
7. Record the results of the measurements on the calibration sheet.
8. The instrument shall be passed only if:
    (i) There is no visible damage to the faces.
    (ii) There is no looseness in the slider.
    (iii) The zero reading is within 1/2 division of 0.00 mm or 0.000 inches as applicable.
    (iv) The two reference readings are within 1/2 division of the actual values of the gauges used.

Source: Nypro Ltd

Metrology

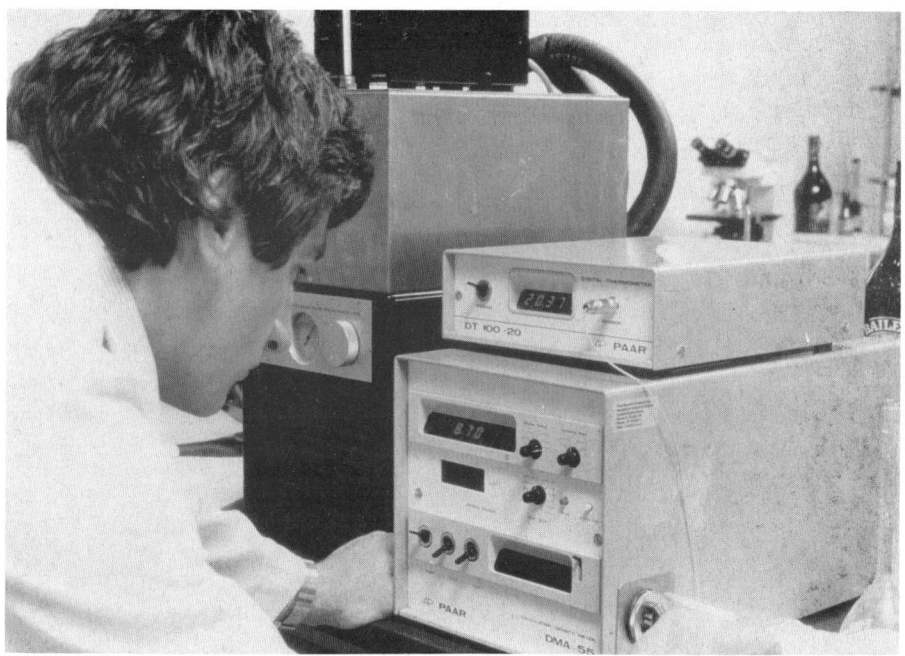

Calibration is essential, even of modern instruments R. & A. Bailey

Errors

An important part of metrology is the understanding of how measurement errors arise. There are five basic types of error.

1. *Instrument error*: This arises from a fault in the instrument being used to make the measurement. The instrument itself may be faulty, as in a wrongly marked measuring beaker, or it may be badly calibrated. The instrument may have problems, especially at the limits of its range, in recording extreme values accurately. Some instruments will produce different values as they go down from a high value compared to those they produce as they come up from a low value. This is called 'hysteresis'.

2. *Personal error*: Arises from a misinterpretation of the instrument, from a misuse of the instrument, or perhaps from an operator's physical problems. Obviously colour-blindness could cause problems, as could short-sightedness or an unsteady hand in interpreting the output of certain measuring instruments.

3. *Systematic error*: The best known example of this kind of error is zero error, where the dial or pointer doesn't start at zero. Any reading taken thereafter will be wrong to the extent that the pointer missed zero. If a barometer wasn't

adjusted to the right height above sea level, all the pressure readings will be wrong.

4. *Random error*: The fourth and most important kind of error is called random error. When any observer makes repeated observations of the same quantity, some variation in the results will occur, no matter how careful the operator is. Why these errors occur no one knows, but scientists have been aware of them for centuries. Indeed the so-called 'normal curve', which is the basis of so many sampling schemes (see Chapter 21), was originally devised to explain the distribution of errors in astronomical observations.

5. *Sampling error*: Another major cause of error is that due to sampling. Clearly if we wish to test the proportion of sugar in a production batch of soft drink, not only do we have to be sure that the hydrometer is scaled and read correctly, but also that the test batch is in fact a fair sample of the main production.

The purpose of metrology is to consider the ways of expressing both states of matter and changes of state in numbers. As the famous scientist Lord Kelvin (after whom the absolute degrees of temperature are named) said, 'When you can measure what you are speaking about and express it in numbers you know something about it; but when you cannot measure it, when you cannot express it in numbers, your knowledge of it is of a meagre and unsatisfactory kind.'

Part of metrology is the development of systems of units and clear standards for them. But the concept of standards has been taken much further and now covers many aspects of the ordinary industrial scene.

Chapter 19

Standards

COMMON standards of measurement (particularly of weights and measures) have been recognised as essential to commercial life wherever trading sprang up in the world. But the Industrial Revolution was more than a hundred years old before the idea of standardisation was extended to engineering and industry generally. As late as 1880, it was a matter of luck if the candle you bought in the grocer's fitted neatly into the candlestick the ironmonger sold you!

As it happened, that didn't matter much, since candles could very easily be trimmed to fit, and anyway they were on their way out. But when British industrialists realised that steel girders were being imported into England from Belgium and Germany because British steel mills 'had too much individualism', and had failed to provide a predictably standardised product, things had to change.

The first meetings of the Engineering Standards Committee in 1901 reduced the number of standard steel sections from more than 175 to 113 and the number of tramway rail styles from 75 to a mere 5. These decisions were estimated at that time to have saved £1 million a year. Since that time, the British Standards Institute (BSI, as it came to be called) has issued hundreds of standards on different matters. It is still technically an independent, industry-maintained organisation, not a government body. As a result of Britain's industrial prestige in the early part of the century, it has served as the model for national standards institutes all over the world.

Importance of standards

A central theme of this book is the importance of the systems approach. The quality approach to making products differs from the craft approach and from the 'human robot' approach of the 1880–1950 era by stressing both the systematic attitude to quality production and the individual worker's responsibility. A key aspect of the systems approach is standardisation. Precise standards are like a clear, unmoving target; the more standards are laid down in detail, the better the production people can develop their skills and techniques. It is also better for the customer, whether industrial or consumer, to know precisely what he or she is getting, and therefore to be able to guarantee its fitness for use.

The products for which it is most difficult to prescribe and maintain quality standards are those dependent on some naturally varying input or process.

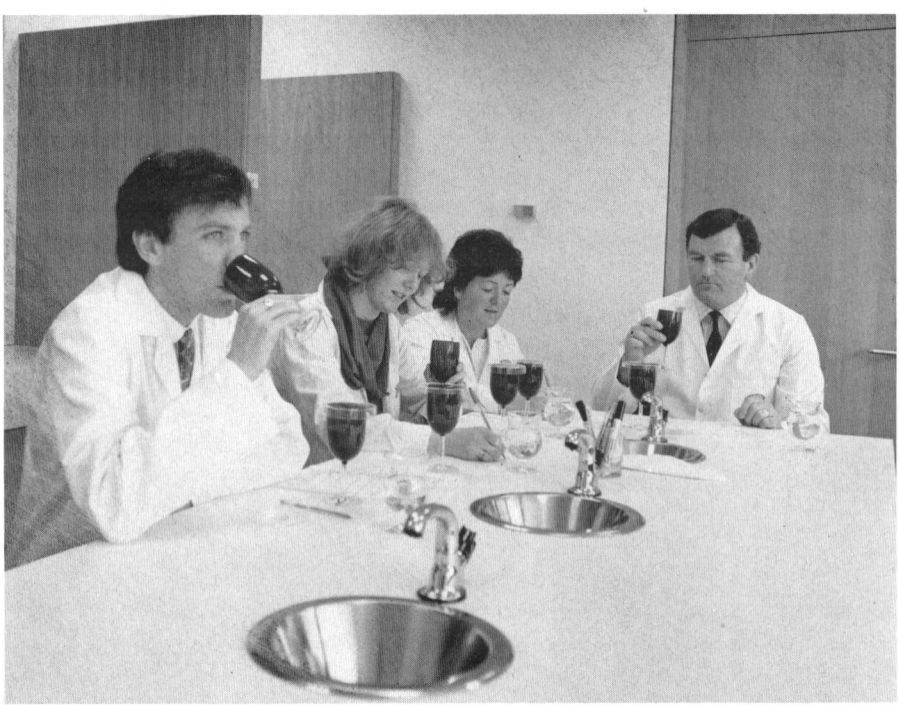

The taste of a natural product must be continuously monitored   R. & A. Bailey

Immense efforts are made by manufacturers of brand goods such as tea or coffee to blend the raw materials so that the final product remains uniform. This effort is necessary because tea leaves, being a natural product, vary subtly in flavour from crop to crop. The makers of wine have even turned this variation to their advantage; people are encouraged to compare vintages. But most ordinary drinkers would prefer to be able to order a standard dry or sweet wine, and know exactly what they were going to get.

The development of standards is therefore a key aspect of the drive for better quality. As the great Japanese quality expert Kaoru Ishikawa (who invented the approach to problem solving discussed in Chapter 20) wrote:

> None of the Japanese success in quality would have been possible if we had not linked progress in quality control with progress in standardisation: they are as inseparable as the two wheels of a cart ... When we want to achieve the statistical control of a production process, we must establish two kinds of standards: standards for materials such as products, parts and raw materials; and standards for systems such as operation, engineering and new product development.

The importance of standards and the practice of standardisation have been recognised internationally. Virtually every country in the world has a standards authority; in Ireland the National Standards Authority of Ireland (NSAI) took over the responsibility from the IIRS in 1985. There are also international standards bodies, such as the International Organization for Standardization (ISO), which works very closely with the International Electrotechnical Commission (IEC). The ISO, which was founded in 1947, has fifty-five member countries. The IEC, founded in 1906, has forty-three. Other international bodies include the European Committee for Standardisation (CEN), its electrical counterpart CENELEC, and the Asian, African and pan-American standards organisations. The defence and nuclear industries have particularly elaborate international standards organisations.

Definition of a standard

The most widely used definition of a standard is that originally promulgated by the UN and adopted by the ISO in 1976. It is also the definition used by the BSI. A standard is defined as: 'A technical specification or other document available to the public drawn up with the cooperation and consensus or general approval of all interests affected by it, based on the consolidated results of science, technology and experience, aimed at the promotion of optimum community benefits and approved by a body recognised on the national, regional or international level.' This is an extremely wide definition, because a standard might cover any one of the following five aspects:
1. terminology, symbols;
2. classification;
3. methods of measuring, testing, analysing, sampling, etc.; methods of declaring, specifying, etc.;
4. specifications for materials or products: dimensions, performance, safety, etc.; specifications for processes, practices, systems, etc.;
5. recommendations on product or process applications: codes of practice.
To be successful, a standard must be:
— *wanted*: the object of the standardisation procedure is to produce economic benefits for everyone by setting up a common specification. Unless this common specification is actually needed, the standard might as well not be drawn up.
— *used*: most standards are voluntary, but there is no point in compiling a set of standards that most people ignore. The economic effectiveness of a standard depends on how many people use it — the more, the better. In some cases, especially in the safety and hygiene area, standards are legally enforceable.
— *planned*: the production of a standard must be part of a long-term plan. Because it is necessary to carry the bulk of the industry with a standard, the first draft may bear the signs of compromise. At the same time, especially in a fast developing industry, it is important not to inhibit development by premature standardisation. For instance, the current IS 300: 1984 'Quality System Management' specifically refers to documents, rather than the computer records and databases that the factory of the future will be managed by. The review and development procedure

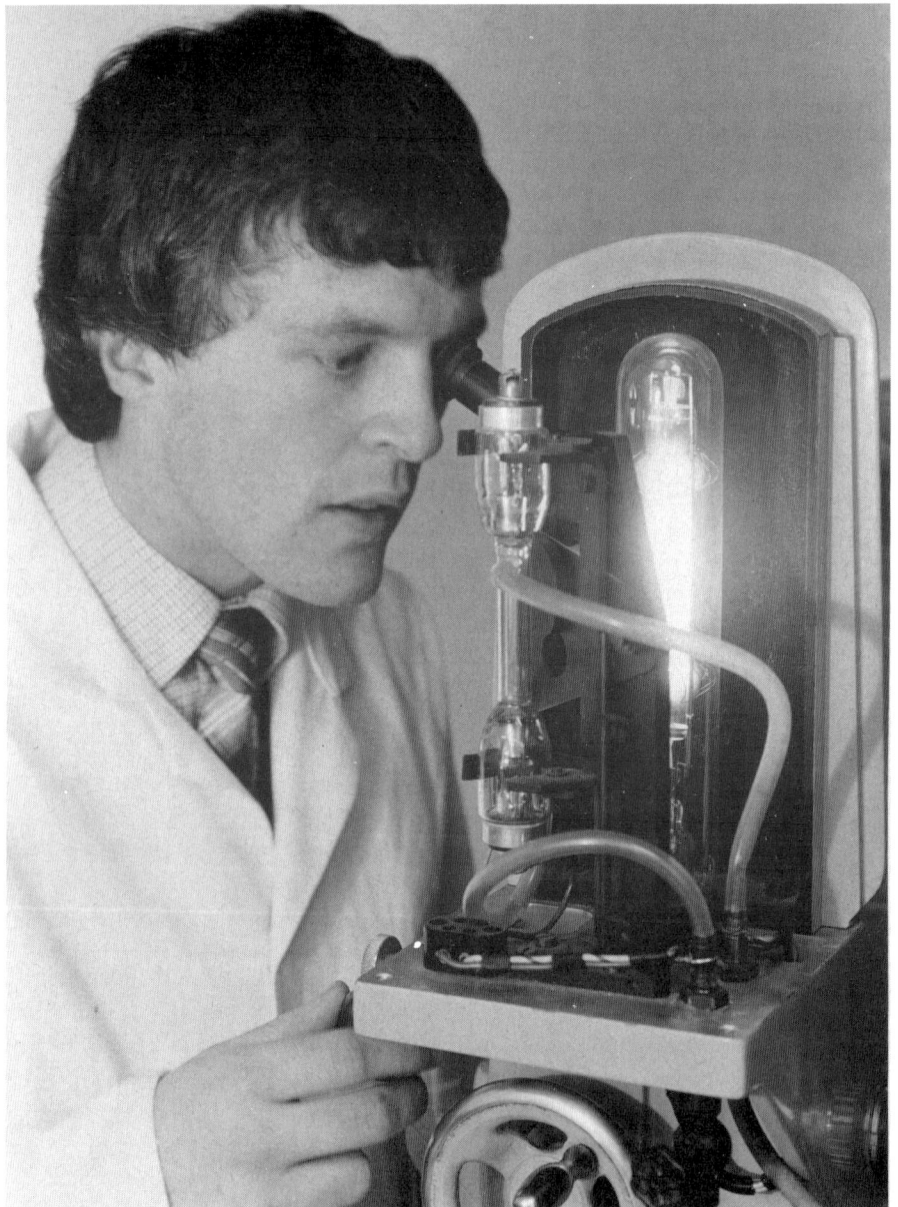

Testing for conformance to standards — IIRS

that ensures that the standard keeps up with the industry is as important as the initial writing task.
— *harmonised*: every standard should be part of a structure of standards, stretching as far as possible through the product chain and across countries. There is no

point in designing a standard candlestick and a standard candle mould if the one isn't related to the other! Also there is no need to re-invent the wheel — if a satisfactory international standard already exists, codified or *de facto*, there is no point in spending valuable local time and money creating a special Irish standard.

Approval marks

The standards concept has been developed by the granting of approval marks to products that have reached certain standards. Typically approval marking is associated with the safety of products, so they are most frequently found on electrical goods, but they also appear on safety helmets, safety glass in cars, etc.

In Ireland the IIRS, through its subcommittee the NSAI, operates a 'Standard Mark' scheme for a number of products, including plastic pipes, concrete pipes, fish, and school stationery. The British Standards Kite Mark is the most familiar of foreign approval marks, though there are a number of internationally organised marks.

A typical national approvals scheme involves the setting of standards, a testing laboratory, a committee of independent experts and a general supervisory board. In Ireland these functions are carried out by the IIRS.

Standards in quality control

Standardisation affects the quality plan in two ways.

Firstly, national and international standards help the manufacturers develop their own quality manual standards. Sometimes these can be taken straight from the national standard. But Ishikawa warns that manufacturers have to develop beyond national standards, which are a compromise and a minimum requirement rather than necessarily a statement of best practice. He tells the story of a paper manufacturer making half-tonne reels of paper for newspapers. The customers complained that the reels kept breaking, despite the fact that they matched up to the Japanese Industrial Standard. Some reels that didn't match the Standard were quite satisfactory. Ishikawa comments, 'This reveals that characteristics in specifications for products and raw materials did not conform to the true characteristics actually wanted by the customer and therefore did not address fitness for use.'

The second impact of standards on the quality plan is from the quality systems standards themselves. At the moment there are seven such standards published in Ireland by the NSAI. These are:

— *Quality Glossary*: IS 257: 1983, which contains definitions of standard quality terms

— *Quality System Management*: IS 300: 1984, which is published in four parts, contains the basic requirements for the introduction and administration of quality systems in manufacturing and service industries

— *Sampling Procedures and Tables*: IS 301: 1984, which provides guidance on the selection and use of sampling plans and procedures for inspection by attributes

— *Quality Manual Preparation Guide*: IS 302: 1984, which provides guidelines for the preparation of a documented quality system

— *Quality Audit*: IS 303: 1984, which defines the principles and methods for the appraisal of a company's own quality system or that of a supplier

— *Guide to Quality Costs*: IS 304: 1984, which provides advice on the identification and control of quality related costs

— *Guide to Quality Management in the Service Industries*: IS 305: 1985, which deals with the special problems of delivering quality in a service environment.

These standards fall into two sets: the standards and guides to specific aspects of the quality programme, such as those relating to the preparation of the quality manual, sampling tables and the quality glossary, and the main overall systems standard, IS 300: 1984 *Quality System Management*.

This standard provides the frame inside which the others operate. It deserves detailed discussion.

The Irish Standard for quality system management

The origin of quality management standards lies in the demands of the buyers of military hardware. As weaponry became more sophisticated, the needs for fail-safe quality standards became more obvious. The NATO armies were the first to codify their quality standards, followed by the American and British ministries of defence, and then other government departments, for their general supplier assessment schemes. Increasingly these standards began to spread — first to the nuclear industry and then through the whole industrial system. Now they are an essential part of the infrastructure of any industrialised country. The Irish Standard is the result of work by a committee set up in 1981 by the IIRS and the Irish Quality Control Association.

The standard is divided into four parts, consisting of an overview and general management specification, and three further specifications covering the requirements for various categories of manufacturing firm. The quality system requirements for a service environment are covered in IS 305: 1985.

IS 300: 1984 Part 0: Quality System Management

The purpose of this Standard is to provide manufacturers with a framework for their own quality systems. At the same time it provides a set of criteria by which suppliers can be judged. Compliance with these requirements is necessary for all firms seeking the standard marks administered by the NSAI, whether they supply products or services. The requirements of the Standard are:

1. *Management responsibilities*: The basic responsibility of management is to see that a quality plan, and the resulting quality policy, quality manual and quality problem analysis systems are in place and operating correctly.

2. *The quality system*: A fully documented quality system must be established which has effect right through the plant's operation from the design and contract stage to the final delivery and after-service.

3. *Contract review*: adequate systems must be in operation to ensure that the customers' needs are clearly stated and that they can in fact be met by the contractor.

4. *Design, development and specification control*: Bad quality starts with bad design, so it is necessary to have all the documentation relating to new designs properly evaluated and controlled against the objectives of the quality plan.

5. *Documentation control*: Proper documentation is the lifeblood of a quality system; management must ensure that all documents relating to the productive operation of the plant are clear, readily available, and up to date.

6. *Purchasing*: Up to 65 per cent of defects in the marketplace can be traced to faults in bought-in products. It is vital to ensure that purchasing is done only from approved contractors who have been assessed for the ability to supply the right quality (see IS 303: 1984 *Quality Audit*). Some form of goods-inwards inspection should be in place.

7. *Item identification and traceability*: Where possible, items used in the process should be traceable, so that any source of fault can be quickly identified.

8. *Production process control*: The production process must be controlled and monitored. Each significant product or service characteristic should be identified, and the inherent variability in the system identified. Process control charts should be used where possible.

9. *Inspection and test*: An inspection and test plan should be drawn up for the product. This should specify the product characteristics and the inspection points relating to them. Both operator and quality control inspection should be specified.

10. *Control of measuring and test equipment*: Test equipment used to check the production process must be adequately calibrated and controlled. The techniques and routines for this should be a special aspect of the quality manual.

11. *Handling, storage, delivery and installation*: The handling system must protect

the product from damage during all phases from raw material to delivery into the customer's hands. The quality manual should prescribe the correct conditions for each state of each product.

12. *Inspection and test status*: The results of inspection tests should be readily available for all stages of manufacture. Route cards, labels, stickers, inspection records, etc. are all useful in this.

13. *Control of non-conforming items*: Rejected or non-conforming products should be clearly identified and a formal procedure laid down as to their ultimate fate. Reworked materials should go through testing again.

14. *Corrective action*: The quality system should outline procedures and people to identify and correct faults.

15. *Quality records*: Records should be maintained to demonstrate the effective operation of the quality system, including the results of vendor assessment procedures.

16. *Quality audits*: A plan of periodic quality audits should be prepared by management. These should be designed to check how well the quality system is working, how adequate it is and how far the products actually meet with the quality requirements of the customers.

17. *Training*: The quality system should include a training programme to ensure that the staff are trained and motivated to meet the quality requirements of the customer.

18. *Marketing and servicing*: The quality system must see the product right through to the customer's hands. Levels of stocks and spares and a complaints procedure are also part of the quality system.

19. *Number and frequency of assessments/inspections and tests performed*: Correct statistical sampling procedures should be part of the test and inspection system.

Parts 1-3 IS 300: 1984 – Quality system requirements

The rest of IS 300 is concerned with setting these general requirements for a quality system into specific industrial environments. Part 1 gives the requirements for a quality system in situations where the specified (technical) requirements of the product or service are stated principally in functional terms. In this case the purchaser wants a product that will perform a certain task, and it is up to the manufacturer to design and produce that product. The standard is therefore primarily concerned with systems that will demonstrate the manufacturer's capabilities in design, development, production, installation and servicing.

Part 2 covers the case of a manufacturer who has to meet the requirements of an already established design. In this case the Standard is mainly concerned with inspection and testing during manufacture. Installation skills can also be important in this category.

Part 3, like Part 2, is concerned with meeting an established design. In this case, however, conformance to specified requirements can be established by inspection and tests conducted on the completed product.

Conclusion

The regulations of IS 300 (covering the same ground as the British Standard BS 5750 and the American ANSI N 45.2) provide the basic framework for the establishment of a quality system. The other standards fill out some of the detail in such matters as exact definitions of words, statistical sampling tables and so on. The standards have been collected together by the NSAI into a *Quality Assurance Handbook*, which presents the various standards in one set.

Chapter 20

Problem solving

SOLVING problems requires both information and insight. Like everything else, it is 99 per cent perspiration and 1 per cent inspiration. Information can be gathered and organised by the techniques described in this chapter, but insight is harder to marshal. To some extent techniques such as 80/20 analysis and brainstorming can help, but fundamentally it needs a creative attack on the situation. Before the exercise starts, some person or group must assume responsibility for the solution — unlike the lady in the Thurber cartoon who snarled down the telephone, 'Well, if I called a wrong number, why did you answer the phone?' Clearly if everyone simply says, 'That's not my problem,' no solution will be forthcoming. Taking responsibility is not the same as taking blame. Responsibility simply means being determined to see the problem through to a solution.

Understanding the problem

The first step is to locate and analyse the problem. For instance, suppose there is a sudden jump in the number of rejections of goods by inspectors. This is wasteful and is hitting production output. For practical reasons, the first thing to do is to consider interim action to reduce the impact of the problem. In this case the obvious first response is to allow more overtime working. Note that the only thing the interim solution does is to buy time so that the real problem can be approached properly. It is not part of the problem solving sequence as such. Usually it treats the symptoms, not the disease.

All our efforts are now devoted to finding the cause of the problem. This should be done in steps:

1. *The problem*: What exactly are we worrying about? In what respect does the actual depart from what should occur?

2. *The symptoms*: This is the problem described in detail. Is it, for instance, all sizes of Model A or only some? The questions to ask are:
— *What?* What are the items involved, sizes, what specifically is wrong with them?
— *Where?* Where on the item is the fault occurring? Or, perhaps it may be more helpful to ask, where in the plant is the fault being discovered, and (if different) being created?

— *When?* Are the faults related to time? Are they produced mainly at the beginning or at the end of a shift? Or at the beginning or end of the week? When were the faults first noticed?

— *Why?* Why is this defect a problem to anyone? Usually this will be obvious, but occasionally this question will light a solution route.

— *Who?* A problem is not a problem in a vacuum. It is always a problem *for* someone. If you can't find anyone for whom an apparent problem exists, then you don't have a problem. Conversely, if you find out for whom the problem matters, you have a possible solution route.

— *How many?* How many of the items are suffering from this problem? How much of the item is affected?

At this stage the important thing is to identify clearly what is part of the problem and what is not. This enables you to relate the cause precisely to the problem, and to test possible solutions.

3. *The causes*: The first assumption is that there is likely to be several causes, contributing to the problem in differing degrees. The next thing is to be sure that the various causes specified are right. For instance a cause that appears to explain the problem might also raise the question as to why the same cause doesn't produce that effect elsewhere. The use of brainstorming and Ishikawa diagrams can very often help in adducing causes. When they are identified, they should be checked to confirm that if the supposed cause is changed, the effect will change as desired. As one writer put it: 'the source of the weakness must be discovered and repaired. A person falling down a manhole is rarely helped by making it possible for him to fall faster or more efficiently.' As far as possible ideas should be tested before the solution is made to depend on them.

Reaching solutions

This is where the 1 per cent of inspiration comes in. The work up to now has been in analysing the problem in increasing detail. Very often a solution will appear in the course of that analysis. The only question then remaining is whether it is the best solution. A solution already proposed has a way of seeming inevitable; its mere existence seems to reduce the enthusiasm for a possible better answer to the problem. One way to counteract this is to make a list of various objectives that a solution must meet. Some of these will be absolute musts, some merely desirables. Different solutions will usually supply the absolute and the desirable outcome in different ratios. The fact that the original solution inevitably doesn't supply every want can stimulate the group to further search.

If a solution hasn't appeared, one must be created. The best approach is to stimulate the group into looking at the problem from several angles, avoiding being led down a particular path by preconceptions. After all, the reason you still have a problem is because the accepted wisdom didn't come up with a solution.

If you put a bee in an open glass bottle, with the bottom towards the sun, it will buzz energetically all day trying to bat its way through the glass. If you put a fly in the same bottle, it skits about in a random fashion and very quickly finds the opening. By its own standards, of course, the bee is right. Such a thing as glass shouldn't exist. But the fly, who works in an apparently chaotic and unintelligent fashion, ignores what should or should not be the case. It works with what is, discovered by trial and error. Various questions can help to free the mind from accepted notions:

— Have we seen this problem before, perhaps in a different form?
— Have we solved a related problem? Or a more general problem, or a more special problem?
— Can we restate the problem in different terms?
— Could we solve the problem if one of the 'musts' was removed? Can we re-examine the list of desirable outcomes?
— Can we solve part of the problem — 80 per cent, say?
— If we assume that a part was solved, perhaps by the introduction of an extra process, can we solve the other part?
— Can we learn anything from the extreme cases?
— Re-examine for whom and why the problem matters — can we do the same thing a different way?
— Are there any special circumstances that might reveal something, such as coincidences?
— Can we learn anything from the inverse of the problem?

The solution produced by these methods must be tested. Assuming that the solution supplies all the 'must' requirements and as many as possible of the desirable outcomes in the short term, we must confirm that there are no undesirable consequences in the long term. Finally, there is the feedback stage. Can you confirm that the problem is in fact solved? Did your plan work out as expected? If not, why not? Keep the records for the next time.

The seven tools of problem solving

Writers on quality control have devised seven diagnostic tools to help in the understanding of quality problems. They were originally devised to help quality circles in their operation, but the same techniques can be used for any problem analysis. The seven tools are:

1. check sheets
2. cause and effect charts
3. stratification
4. 80/20 analysis
5. scatter diagrams
6. histograms
7. control charts

Problem solving

1. *Check sheets*: Check sheets are charts to help in the location of problems. These are laid out so that the number of incidents of a problem can be located area by area. For instance, the customer relations department of a car manufacturer received several complaints about body rust. The exact spot of the rust was marked on an illustration of the car for each complaint. Very quickly it became obvious that most of the complaints related to the lower left door post panel. Obviously there was a design or quality problem in that area.

The purpose of check sheets is to gather information in a systematic way. This enables the problem to move from a generalised complaint — 'There is too much rusting on the body of this model' — to a solvable problem.

There are various types of check sheet, depending on the problem to be solved. For instance, the company may wish to identify the causes of rejected work. It will set up a Defective Item check sheet. This lists the main possible causes of rejection — such as scratches, dirt, defective parts, wrong size, etc. Every case of each cause is marked on the check sheet, and soon a picture emerges as to the most prevalent type of defect.

The next step is to take that defect and list the places or machines that could have caused it. This is the Defective Location check sheet. It might consist of a plan of the plant, with the source of defects marked as they occur. As information builds up, the picture of the cause and location of defective work becomes clear.

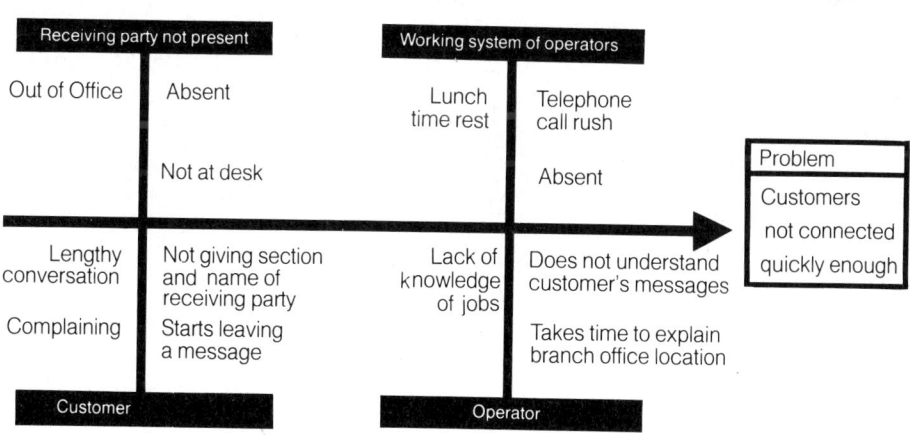

A cause and effect chart

The most frequent source of that defect can then perhaps be further explored. This involves using a Defective Cause check sheet. Each possible cause is listed, and the number of each recorded. Eventually, by a process of elimination, the precise source of the defect is identified. If there is more than one cause then each should be tackled in turn.

The check sheet idea comes in many shapes. Every factory and every process will have its own design. Their purpose is to enable the data to tell the problem solver more about the problem.

2. Cause and effect charts: One of the most powerful tools of problem solving is the cause and effect chart, otherwise called the fishbone or Ishikawa chart, after the Japanese quality expert who developed it.

The theory of the chart is based on the idea that no human event has a single cause. In ordinary life, this is easy to accept. Road accidents are typically caused by careless driving *and* bad conditions; strikes are a function of bad management *and* aggrieved workers; it takes two sides to make a quarrel.

The cause and effect chart is a way of displaying all the contributory causes to a single event or effect. On the extreme right of the figure is the effect. Lines flowing into the effect show the contributory causes. These are usually categorised as main causes and sub-causes, and perhaps sub-sub-causes. The final chart should show the whole range of contributory causes to the single event.

The first step in creating a cause and effect chart is to *identify the effect*. This may sound unnecessary. But if the exact effect isn't clearly described, the identification and evaluation of causes will be hampered.

The second step is to *construct the framework* of main causes. This is usually best done by identifying the inputs to the process that is being examined. Thus for a simple production process, the bones of the fishbone chart might be materials, people, equipment and management. A more elaborate set of causes might work instead by identifying the major sub-processes in the production line.

The next stage is the *brainstorming*. All the members of the problem solving team are encouraged to identify possible sources of the problem, under the main categories. Someone should be delegated to write down the sub-causes in a network on a blackboard or flipchart. In this way ideas are placed in a logical relationship to each other.

At this stage there should be no discussion of the causes, as this would inhibit the free flow of ideas. The purpose is to identify all the possible sources of the problem, however unlikely. In fact the craziest ideas may possibly stimulate someone else in the group to produce a really valuable contribution. The brainstorming session should be long enough for people to contribute their own ideas, and then to develop other people's. At the beginning of the session everybody is concerned with their own ideas; only when the brain is empty of the first thoughts can the really creative interaction take place.

Don't worry if the chart is very lopsided at this stage. It can be tidied up later.

This may happen, for instance, if one of the main causes, for instance materials, is found to contribute nothing to the problem, while another, say equipment, needs to be examined in great detail. The final stage of the brainstorm session is to leave the blackboard or flipchart in place for a week or two, to allow extra thoughts to develop.

When the layout of the chart is complete, the team must ascribe *relative importance* to the various causes. This will have been partly done during the brainstorming session, since piling up ideas in one corner of the chart will direct attention to that cause. On the other hand, a simple cause such as the variability of the raw materials may in fact be more significant.

The purpose of the cause and effect diagram is to increase the depth of analysis of a problem. By brainstorming as many possible causes for a single effect, the problem solving team can be sure they haven't missed anything important. They can then confidently sit down to the job of evaluating the relative importance of the causes.

3. *Stratification*: This is the technique of classifying existing facts into groups with common characteristics. It aims to select causes from single factors. As such it is an essential part of any problem-solving process. Good stratification can often, by itself, reveal the cause of a problem. In a normal production line, several machines are involved simultaneously producing the same product. There may also be several shifts, and even possibly several factories. If the quality staff discovers a problem with non-conformance, they have to work towards locating the source of the problem by dividing the sources of output into *strata* (a Latin word meaning layers). The technique of stratification is in fact just like the old Roman military tactic of divide and rule.

As soon as the quality people spot the problem, they have to find the source as quickly as possible. To do this they will divide the various sources of output into sets, and examine the output of each set. This can be done in various dimensions. They might start with the theory that the fault is a general one, for instance that all the machines tend to run out of true at the end of a shift, or they may believe that one particular machine is out of true.

In both cases the researcher will have to split the products into groups. Following the first theory, the products would be split by time of production. The products produced during each hour of the shift would be separated, so that quality assurance staff can assess whether there is any increase in non-conformance as the shift progresses. The second theory requires a split by machine. In this case the output of each machine would be separated, and the proportion of non-conforming output checked. Stratification is a technique for breaking down a mass of data into manageable groups.

4. *80/20 analysis*: 80/20 analysis is the name given to a remarkable discovery. In the nineteenth century an Italian economist called Vilfredo Pareto noticed

that 20 per cent of the people in the economy own about 80 per cent of the wealth. What Pareto didn't notice was that many other relationships follow the 80/20 rule, which is often called after Pareto. For instance:
— 20 per cent of drivers cause 80 per cent of motor accidents;
— 20 per cent of donations to charity make up over 80 per cent of the total cash received;
— 20 per cent of products in a department store make up 80 per cent of sales;
— 20 per cent of people make 80 per cent of telephone calls.
No doubt there are many examples of this rule in operation in every company.

Of course the relationship isn't necessarily exactly 80/20. In fact in Ireland less than 10 per cent of the people own 75 per cent of the wealth. Similarly a management consultant's study of an Irish book distribution warehouse discovered that 13 per cent of customers accounted for 71 per cent of sales. The crucial point is that the categories divide into the vital few and the trivial many. In quality management things work exactly in that way. The majority of defects or non-conformance problems can be traced to a few major causes. A typical result is that discovered by the Westinghouse (WESL Ltd) factory in Shannon, where they discovered that the top ten faults accounted for 82 per cent of the total, and the top five for 68 per cent. The quickest way to solve the bulk of a problem, therefore, is to discover and cure the vital few.

The 80/20 diagram is a way of grouping data so that the expected relationship of the vital few and the trivial many is revealed. The first step is to list all the major causes of a problem. Just as in stratification, this might be by machine, by shift, by cost, by operator, or by any other factor. The data should then be collected under these categories and summarised.

Exhibit: Shortening customers' telephone waiting time
A quality circle in a large American bank was presented with the problem of preventing customers from getting irritated if their call was not answered immediately. An average of 500 customers called the bank every day. Surveys showed that they didn't like to have to wait for more than five rings before being answered. The company believed that every call should be answered after two rings. The problem was to discover why customers were kept waiting.

The first approach was to create a cause and effect chart, listing the various reasons for delays. The major reasons were
— receiving party not in his/her office
— not enough operators
— operator didn't understand enough about the company to put the customer through to the right place
— the customer failed to explain the problem.
The various sub-causes were detailed on the fishbone chart (see p. 169). Each of these main causes was then put on to a check sheet, and the operators asked to tick every occasion under each cause for twelve days. The totals revealed that nearly thirty customers or potential customers were being kept waiting every day. Finally these figures were summarised in an 80/20 analysis.

80/20 Analysis of reasons for waiting

Reason	Total	Per cent
One operator missing	172	49
Receiving party not present	73	21
No one in relevant section	61	18
Section not given by caller	19	5
Other reasons	26	7
Total	351	100

The top three reasons accounted for 88 per cent of all delays, and in fact the second and third reasons were two aspects of the same problem of there being no one to receive the call. The quality circle produced two solutions to the problem. The first was a change in the operators' shift system, so that there would always be two operators on duty, and the second was a more problematic campaign to encourage people leaving their desks to leave notes saying where they had gone. The follow-up study showed that the change in the operators' shift reduced the delays from this cause from 172 in twelve days to 15. Overall, delays of this sort were reduced from 351 to 59; 63 per cent of the new figure was due to people not being at their desks.

One cause or group of causes will most likely spring to the eye as contributing the bulk of problems. Draw a bar chart, with the elements in order of significance, with the most important on the left and so on down to the least important factor. A cumulative percentage line can also be drawn if desired.

The purpose of 80/20 diagrams and analysis is to identify problem areas and

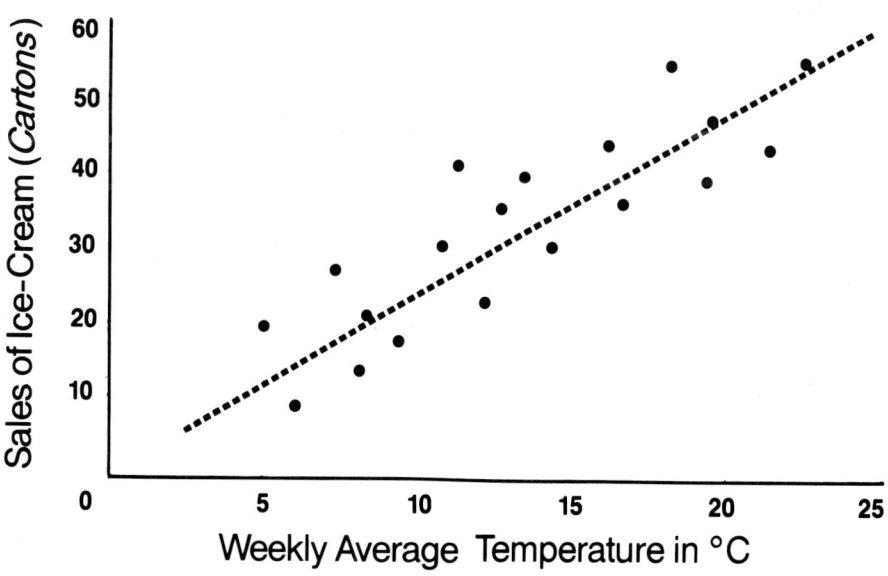

A scatter diagram relating the sales of ice-cream to average daily temperature

to clarify the priority of the elements. The central theory, that many relationships can be divided into the vital few and the trivial many, however, has a much wider application than merely in problem solving.

5. *Scatter diagrams*: Scatter diagrams are used to decide whether two factors are related, and if so what the relationship is. For instance, if we relate the weekly sales of ice-cream to the average temperature, it comes as no surprise to find that the higher the temperature the more ice-cream is sold. It is even possible to use this relationship to predict, fairly accurately, the future sales of ice-cream, given average temperatures for the time of year.

To discover whether this kind of relationship exists, draw a chart with the temperature in degrees along the bottom (the horizontal axis) and sales of ice-cream on the vertical axis. It is a convention of these diagrams that the supposed cause goes on the horizontal axis. The cause is also called the independent variable — the sales of ice-cream certainly don't affect the weather, whatever about the other way around. The effect (which statisticians call the dependent variable) always goes on the vertical axis.

As the figures for temperature and ice-cream sales come in week after week, they are plotted on the diagram. If the only element affecting the sales of ice-cream was the temperature, the points would fall in a neat line rising to the right-hand side of the page. But, as Dr Ishikawa says, nothing is simple in human affairs. Sales rise and fall unpredictably. Perhaps a bank holiday might increase them, or perhaps they jump in May, regardless of temperature, as so many children celebrate their First Communion then.

However, despite these 'blips', the scatter diagram does reveal a clear relationship — the higher the temperature, the more ice-cream is sold. Clearly the technique can easily be used to analyse whether certain factors are or are not responsible for quality problems. The danger of scatter diagrams is that they can appear to show such a relationship whether one exists or not. In the case of ice-cream we can be confident that there is a true cause and effect relationship operating. We are confident in our theory that people are likely to eat more ice-cream in hot weather. But the fact that an apparent relationship has been demonstrated by no means proves that the effect is a direct result of the apparent cause. In one study, a facetious statistician demonstrated an apparently strong relationship over the years between the birth rate in a Norwegian village and the number of storks nesting nearby!

The relationship may not be as simple as the straight line dependency between temperature and ice-cream sale. If you put a pan of water on the stove, the temperature of the water will increase steadily according to the length of time on the heat. But once the temperature hits 100°C, the relationship will cease to hold. However long you leave the pan on the stove, the water will get no hotter. The relationship between heating time and water temperature that seemed straightforward between 60° and 70° and so on has suddenly ceased to work.

In other cases the relationship may take the form of a curved line, or perhaps the scatter diagram will reveal no apparent relationship at all.

6. *Histograms*: Gathering masses of data about a problem is generally easy; what is not so easy is organising the data so that it reveals something. Histograms, also called bar charts, are one useful way. Essentially, they communicate the underlying 'shape' of a set of figures. Other methods are various types of graphs and pie charts.

The first step in creating a histogram is to organise the data. This is usually done on a tally sheet, using the gate count method. The full possible range of the data (in our example, from 13 mm to 22 mm) is listed down the left-hand side of the page, and each observation is marked with a stroke against the relevant row. Every fifth score on that row is struck through the previous four, thus producing the gate-like look. This method greatly simplifies counting at the end.

Once all the observations have been scored, the totals are written against each level, and the relative frequency calculated (see % column in our example). The shape of the information is beginning to show itself. The next step is to turn these figures into a bar chart. As before, the thing observed goes on the horizontal axis and the number of observations on the vertical axis. Above each observation on the horizontal axis a rectangle is drawn with a height equal to the number of observations. This gives the familiar mountain range shape to the histogram. The important thing to remember is that it is the *areas* not the heights of the rectangles that should be proportionate to the frequency of the observation. If the widths of the observation class are the same (and this is by far the best practice) the height will of course be proportionate as well.

In many cases the histogram will follow a standard bell-shaped pattern, with a high peak and both sides falling gracefully and symmetrically to the outer edges of the chart. So frequently does this pattern occur that it is called the 'normal distribution'. But occasionally different patterns will appear. A two-peaked (bi-modal) distribution suggests that one set of observations results from different conditions to the other. This is the moment to introduce stratification to your problem analysis.

7. *Control charts*: The previous problem solving tools have tended to concentrate on unearthing the source of a problem, and are ways of exploring data generally. The control chart, however, is the best way of examining a process as it is in operation. As a part of normal process control, it is discussed in Chapter 23. In this context, it is considered as a method of observing a process to identify a problem.

Suppose that normal quality assurance inspection reveals that a certain defect is causing a mounting number of rejections. It may not be clear where or when during the shift the non-conformance is occurring. A control chart, linked with a series of special inspections for that defect, will very quickly pinpoint where and when the proportion defective moves outside acceptable levels.

Building up a Histogram
1. The Tally Sheet

mm	No.	Total	%
13	/	1	0.4
14	ℋ ///	8	3.2
15	ℋ ℋ ℋ ℋ ℋ /	26	10.4
16	ℋ ℋ ℋ ℋ ℋ ℋ ℋ ℋ ℋ ℋ	50	20.0
17	ℋ ℋ ℋ ℋ ℋ ℋ ℋ ℋ ℋ ℋ ℋ ℋ ℋ ℋ //	72	28.8
18	ℋ ℋ ℋ ℋ ℋ ℋ ℋ ℋ ℋ ℋ ℋ /	56	22.4
19	ℋ ℋ ℋ ℋ ℋ ℋ	29	11.6
20	ℋ /	6	2.4
21	//	2	.8
22		—	—
		250	100.0

1. A suspect batch of rods was carefully measured and the results marked on a talley sheet.

2. The totals by each length could then be easily found and the relative frequency calculated.

3. The result could then be plotted on a histogram.

2. The Histogram

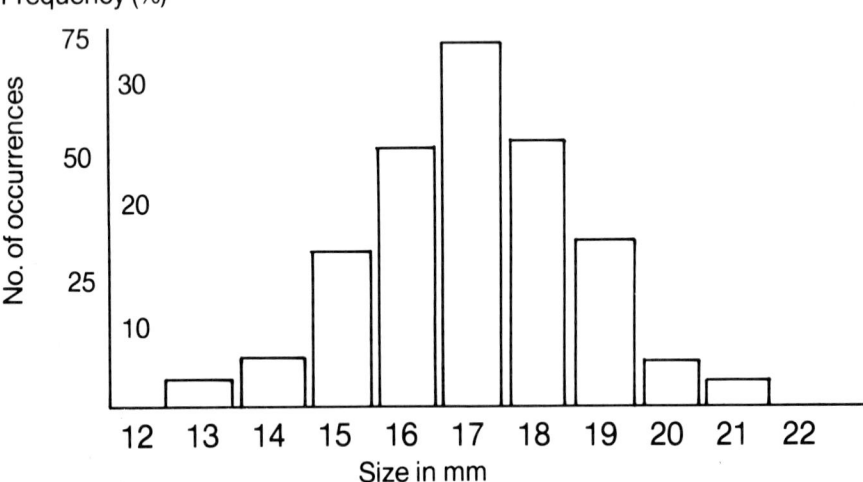

Note that, as expected, the distribution is approximately normal, with a mean of 17.04mm and a standard deviation of 1.4.

Building up a histogram. Note the connection between the number of occurrences and the relative frequency

The chart looks like a normal graph frame, with the sample number or sample time along the horizontal axis and the measured value up the vertical axis. Parallel to the horizontal axis there are usually five marked lines. The centre line is the desired average attainment. This could be a figure such as average measured weight of a sample, or the proportion of a batch with a certain defect. Above and below the centre line are warning lines. If the process is beginning to run out of true, the dots indicating the average value for subsequent samples will appear to drift in the direction of one of the warning lines. This indicates that the process is slipping out of control. Above and below the warning lines are the action lines. The process average should not be allowed to fall outside these lines. Control charts, designed specially to monitor one particular aspect of the process, can be a very valuable way of diagnosing problems.

Conclusion

Problem solving is both a skill and an art. The techniques described in this chapter provide ways of getting closer to the information, and perhaps discovering that crucial information is missing. For instance, the brainstorming session to create a cause and effect chart may reveal that a major set of causes is simply not being adequately monitored. The tools described here mostly developed against a background of the greatest of all information handling techniques, statistics. This discipline is the subject of the next chapter.

Chapter 21

Letting statistics help quality control

THE twentieth century is the era of large numbers. There are more people in the world than ever before, more goods, more money, more of virtually everything (except possibly wildlife). To enable us to understand and manipulate these immense quantities, a new science has evolved: statistics. It sprang from a combination of the theory of probability, invented by aristocratic gamblers in the seventeenth century, with the accumulation of masses of raw numerical information initiated by economists and social scientists in the nineteenth century.

Of course people had collected figures before the nineteenth century. An early form of statistical control chart was in force in London from 1603, when the court and the upper classes protected themselves from outbreaks of plague by having the causes of every death listed once a week. If the number dying from 'fevers, etc.' rose to dangerous levels, Parliament adjourned to the country. During the nineteenth century, however, organisations such as the Royal Dublin Society began to gather statistical information about the country in order that the 'greatest good of the greatest number', which was the political slogan of the day, could be ascertained and delivered.

The first to apply the new statistical methods to quality control were scientists in the Bell Telephone Laboratories in the US. In 1924, Shewhart produced the first version of the famous control charts that are now used all over the world. Over the next ten years the basic charts and sampling schemes were worked out. Quality control managers in industry gradually began to understand how these powerful methods of handling figures could improve quality. The urgent requirements of the Second World War were particularly effective in spreading the use of statistical techniques. So much so that for many years statistical quality control was virtually synonymous with quality control. Today it is one of the most powerful of the quality disciplines.

Statistical ideas should be viewed as a kit of tools which may influence decisions. By combining the controlled accumulation of information by samples with the laws of probability, quality workers can learn much more about the variability and control of manufacturing processes, even than can be learned by 100 per cent inspection.

Summarising information — averages

People quickly discover that large quantities of numbers ('raw data') are difficult

to interpret and compare. The first requirement is to find some way of summarising the information, and of comparing one group with another. We need one figure that can meaningfully represent the whole group. From marine insurance practice the word 'average' was taken. There are three types of average or summary figures. These are:
— the mean
— the median
— the mode.

1. *The mean*: The best known type of average is calculated by adding all the values and dividing by the number of occurrences. This is technically known as the 'arithmetic mean'. To find a team's average number of goals scored per game, for instance, take the total number of goals, and divide by the number of games. In most cases this is a convenient and sufficiently accurate summary of the information. But if we were discussing a set of data that had extreme values, the arithmetic mean would be quite misleading. If my brother is a millionaire, and I am on the dole, the figure for the family's average income is unlikely to represent either his income or anyone else's. A few very large figures in the set will bias the average, which as a result is not an accurate summary of the information. Any set of data which falls into the 80/20 distribution would be unsuitably described by the arithmetic mean. To get over this problem, two other ways of summarising the information are used.

2. *The median*: The median is the value above or below which lie 50 per cent of observations. To find the median first rank the observations in order of value (e.g. 10, 9, 8, 8, 7, 5, 3, 1, 1) and then select the central value (e.g. 7) if there is an odd number of observations or the average between the two central values if there is an even number of observations. The median is less influenced by extreme values than the mean. But because it is a geometrical value — it represents the geometrical centre of the distribution — rather than arithmetical like the average, it is less useful for further mathematical work.

3. *The mode*: The mode is the most often occurring value in a set of data. It is the value represented by the highest bar in a histogram. This is the value meant when a politician refers to 'the average man or woman in the street'. One advantage of the mode is that it is always a real figure. The arithmetic mean of 3.53 children per family makes little sense, whereas the modal family of four children is instantly picturable. One very important use of the mode is in diagnosing data that comes from two disparate sources. If the histogram has two humps the distribution is said to be *bi-modal*. The data in the median example above is bi-modal: both 8 and 1 appear twice and are therefore modes. A graph of the heights of a group of boys and girls would certainly be bi-modal; output

Quality in Practice

from two different machines mixed together can often be bi-modal. In order to develop the analysis, the two sources (strata) have to be separated.

Exhibit: Calculating averages and ranges
A sample from the night production run in the automated widget manufacturing company measured as follows:
11, 10, 9, 8, 8, 7, 6, 5, 3, 2, 2, 1
Quality assurance were asked to calculate:
The arithmetic mean
11 + 10 + 9 + 8 + 8 + 7 + 6 + 5 + 3 + 2 + 2 + 1 = 72
Mean = 72/12 = 6
Median
Because there is an even number of observations, we take the average of the two central figures.
Median = 7 + 6 = 13/2 = 6.5
Mode
Two figures appear twice, 8 and 2. These are both modes, therefore, and the distribution is bi-modal.
Range
The highest observation is 11 and the lowest 1.
Range = 11 − 1 = 10
Semi-interquartile range
Delete the top 25 per cent of observations and take the highest remaining figure, which is 8. Delete the bottom 25 per cent of observations and take the lowest remaining figure, which is 3.
SIQR = 8 − 3 = 5

Summarising information — scatter

The average value, being only one figure out of many, may actually be close to only a minority of all the values, especially in a widely dispersed population. The manufacturer who made shoes only for the average foot size (even if he chose the modal size) wouldn't sell many. There are far more non-average feet than average. We therefore need to be able to summarise the spread of values. The same average can represent very widely different spreads of figures. The mean of the two figures 16, 1 is the same as that of 8, 9. The spread or scatter is very different.

Statisticians distinguish between *precision* and *accuracy*. Consider a bottling line that consistently underfills the bottles by 3 mm. This is precise, but inaccurate. Another bottling line may vary from overfilling to underfilling, but the batch average meets the process specification. This is accurate but imprecise. Usually, a precise but inaccurate process is easier to rectify than an imprecise one. Averages describe accuracy, the measures of scatter discussed below describe precision.

There are two commonly used measures of scatter:
— the range
— the standard deviation.

1. *The range*: Calculated by subtracting the greatest value by the least. In a set of data describing the height of the men in a town, this might be 4 feet 6 inches (the shortest) subtracted from 6 feet 9 inches (the tallest). The range is therefore 2 feet 3 inches. Unfortunately the range suffers from dependence on single extreme values. If, for instance, a circus happened to be passing through town on the day the research was done, the range might have gone from dwarfs at 3 feet to giants at 7 feet, i.e. nearly half as wide a range again. To get over this problem, statisticians use the *interquartile range*. This is calculated by ignoring the highest 25 per cent of values and the lowest 25 per cent of values, and then subtracting the lowest remaining values from the highest remaining values. The disadvantage of the range is that it doesn't consider every item of data, and so is not satisfactory for further arithmetical use. Its advantage is extreme ease of calculation, particularly of ordered data. It is also very easy to understand.

2. *The standard deviation*: Another approach to expressing scatter is to take an arithmetic mean of the divergences from the average. This we could calculate by subtracting the actual figures from the mean and then dividing the sum by the number of figures. We would then have two related figures to describe the data; the average, and the average of the *deviations* from the average.

Unfortunately a straightforward sum of the deviations from the average will always result in zero! An example should make this clear. Consider the sample values 3, 7, 8, 11, 16. Their arithmetic mean is 9 (45 divided by 5). Subtracting 9 from each number and dividing by 5 to get the average deviation, we get $(-6 - 2 - 1 + 2 + 7)/5$. This procedure gives us the unhelpful value of 0 as the average deviation.

To get over this problem, statisticians exploit the fact that the square of a negative or positive figure is always positive. They therefore calculate the *variance*, which is the arithmetic mean of the squared deviations. Squaring the figures in the example above, we get $(36 + 4 + 1 + 4 + 49)/5 = 94/5 = 18.8$. Because quality control work almost always deals with samples rather than whole populations, it is usual to divide the squared deviations by one less than the sample size $(n-1)$ rather than the sample size (n). This is to overcome the fact that samples tend to be biased towards the average. So here the true variance would be expressed as $94/4 = 23.5$.

The variance is widely used in sophisticated statistical research, but it has some disadvantages. Most obviously, because it is an average of squared values, it is out of scale with the rest of the data. If the values are large, the variance can be enormous. For instance the variance of a set of figures expressed in thousands will be in millions.

To overcome this, statisticians take the square root of the variance to produce the *standard deviation*. In the set of sample data above, the variance is 23.5, so the standard deviation is 4.85. The standard deviation and the arithmetic mean are the two most powerful basic figures in statistics. Once they have been calculated for a typically symmetrical, hump-shaped set of data, statistical research has shown that one can assume that roughly two-thirds of the values will fall within plus or minus one standard deviation of the average, and 95 per cent within two. The properties of the normal curve, from which these figures are taken, are central in the development of statistical quality control.

Exhibit: To calculate a standard deviation
Find the mean and standard deviation of sample 11, 8, 7, 6, 8
Step 1: Calculate the mean
11 + 8 + 7 + 6 + 8 = 40
40/5 = 8
Mean = 8
Step 2: Calculate absolute differences of the values from the average (i.e. ignoring signs)
3, 0, 1, 2, 0
Step 3: Calculate the square of these differences
9, 0, 1, 4, 0
Step 4: Calculate the sum of the squares
9 + 0 + 1 + 4 + 0 = 14
Step 5: Divide the sum of squares by the number of items, less one to allow for the bias. This is called the variance.
14/(5 − 1) = 3.5
Step 6: Calculate the square root of the variance, to get the standard deviation.
$\sqrt{3.5}$ = 1.87
This is the long way round. There are short-cut formulae and many calculators and microcomputers will derive a standard deviation automatically.

The proportion

A very important distinction in sets of figures is between those organised into discrete groups or sets, and those which can take any value along a scale. On the weekly shopping list flour, sugar and rice can be bought in any quantity at all; eggs, cakes and wine are sold in prefixed quantities. You can (in theory) buy 1.263 lb of flour, but you can't buy half an egg. The heights of men in our town mentioned above vary from each other in a continuous line from the smallest to the largest. Height, like other natural quantities measured in figures, is a variable.

Figures organised into groups measure attributes: products either conform to specification or not; men are either above six feet or not; defects are either critical, major or minor. This way of handling information is very simple. Once the sets are defined, the values are easy to assign. The data is then summarised by the *proportion*. The proportion is calculated as the number of observations in a specified category divided by the total number of observations. Proportions are very important in certain types of quality control work; 80/20 analysis seeks to define a set of data by a critical proportion.

Probability

So far this chapter has been concerned with methods of describing sets of information. But if we wish to act on what we have learnt, the concept of probability becomes important. What is past and what has been measured is to that extent certain. By adding the ideas of probability, the results from one case can be made generally applicable.

The ideas involved in probability are not easy. Because they are based on the law of large numbers, they often appear to contradict common sense. For instance, after tossing a coin and getting six harps in succession, it is difficult to believe that there isn't a strong probability of the next toss producing a head. Equally, hardly anyone finds it easy to believe that there is a better than 70 per cent chance of two players in a hurling match having the same birthdate. Yet it's true (see the calculation below). The independence of probability from personal action is also difficult to grasp. Take the story of the young man who was terrified of terrorist bombs on plane journeys. He had been told that the chances of there being two bombs on the plane were one in ten million, so he took along his own bomb to protect himself and his fellow passengers.

Probability is measured along a scale of 0 to 1. Nought is absolute impossibility, 1 is absolute certainty. Death and taxes are absolutely certain, human life without water is absolutely impossible. In between these certainties lie probabilities. They are expressed as the proportion of the number of times the measured event occurs to the number of possible times it could occur. Thus if we toss a coin a hundred times we expect to get (roughly) fifty heads and fifty harps. But the law of large numbers considers that even one hundred tosses of a coin is a tiny quantity compared with the infinitely large number of possible tosses. *Even if every single one of the hundred came up harps* there would be no additional expectation, beyond the pre-existing 0.5, of a head. There is no built-in correction device. Because the coin has no memory, the chance is *always* 0.5. Every possible sequence of outcomes has the same chance of occurring, some are just more spectacular than others. In competitive play, however, one might be tempted to check the coin before tossing again.

It's easy enough to see that the chance of throwing a six from a dice is 1/6. The next stage is to discover the chances of throwing two sixes with two dice. The rule here is that the combined probability equals the probabilities of the two individual events multiplied together. The chances of two sixes together are calculated as $1/6 \times 1/6 = 1/36$, or 0.028 in probability terms. The chances of *not* getting two sixes are the reverse of that, i.e. 1 (which represents the sum of all the possible outcomes) $- 0.028 = 0.972$.

When the second event is affected by the first, a more subtle problem arises. For instance if my drawer contains four blue socks and four brown ones, the chances of my picking a blue one in the dark are clearly $4/8 = 1/2 = 0.5$. But I need two blue socks. For my second pick the chances of getting a blue one are less. They are in fact 3/7, since there are now three blue socks and four brown socks left in the drawer. The chances of getting a blue pair under these conditions are therefore 1/2 × 3/7 = 3/14 = 0.21. Obviously the chances of getting a brown pair are the same. But suppose that I didn't mind if I wore brown or blue, just as long as they were a pair? In that the chances are 1/1 that the first pick will be satisfactory, and 3/7 that the second will. The total likelihood therefore is 1/1 × 3/7 = 3/7 = 0.43. The chance of *not* getting a pair is therefore 1 − 0.43 = 0.57.

Exhibit: The birthday paradox
List all the players in the hurling match in alphabetic order. The chance of the first player having a birthday on some day in the year is certain, i.e. 365/365 (for simplicity we can ignore leap years). The chance that the second player has a birthday on some other day is virtually certain, i.e. 364/365. The chance that the third player's birthday is different again is nearly as likely, in fact the chances are 363/365. Similarly with the fourth player, whose chance is 362/365. Every new player has a slightly smaller chance that his birthday will not already have been allocated. Since each of these birthdays is statistically independent, to get the total probability we must multiply the individual probabilities. This gives the following enormous sum for the probability that no birthdays are in common over two teams:
365/365 × 364/365 × 363/365 × 362/365 ... 336/365 = 0.2936837
Since this is the chance that no two players have the same birthdate, the chance that they do is
1 − 0.2936837 = 0.7063163.

In other words there is a 71 per cent chance that two players in a hurling match will have the same birthday.

The true probability that two of the hurling players have the same birthdate is higher than calculated, because births tend to fall into a seasonal pattern. This illustrates an important requirement of the theory of probability. The arithmetic can be correct only if the occurrences of the events are not biased in some way, as in the example of the man with the bomb. The dice must be properly thrown; the selection of socks truly blind. Since 30 per cent of Irish people are under fourteen, we might guess that the probability of the next person you meet being under fourteen is 0.3. Or is it? In fact it almost certainly isn't. The chance of the ordinary working person meeting anyone under fourteen either at work or in normal social intercourse is slight. The way society works and plays stratifies age groups, so that it is actually very difficult for pollsters and market research people to get a good random sample of all ages.

Frequency distribution

In its raw form, each set of data is different. The special circumstances of the data itself and the collection make it so. In order to make predictions and extended use of the data, statisticians try to identify which of certain well known families

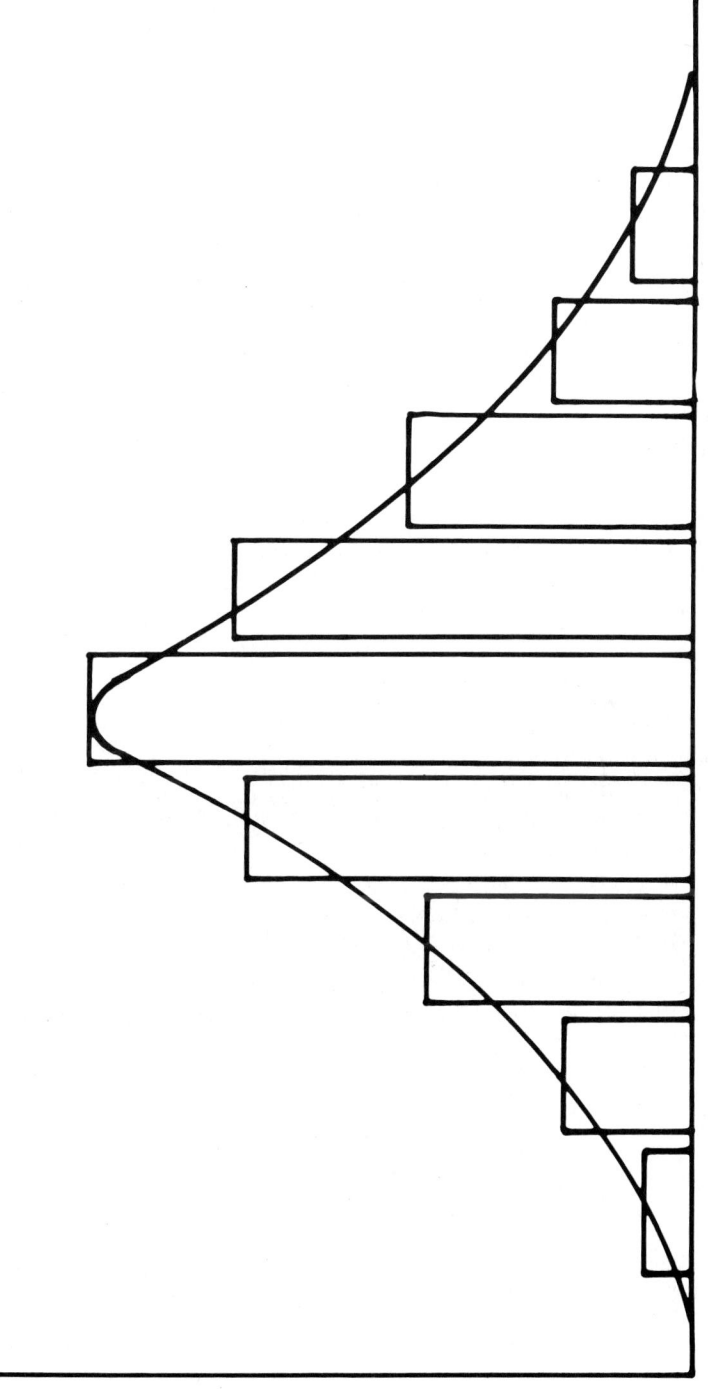

Deriving a curve from a histogram

of shapes the data fits best. The statistician has, as it were, various off-the-peg suits in the cupboard (such as the normal distribution and the 80/20 distribution), which he or she will try to fit to the data. The advantage of these suits is that their properties have been studied in great detail, so a lot is known about them. If our set of data can be made to fit or match one of these distributions, then we can use the known properties of the ready-made model to draw conclusions about our information. The fact that the suit usually won't fit exactly rarely matters. Normally the inferences drawn from a set of data are sufficiently robust (i.e. require only a certain level of exactitude) that it doesn't matter that the distribution isn't an exact match to the mathematical ideal.

The first step in finding a fit is to transform the data from the step-structured histogram to a smooth curve. If the curve is drawn properly, the area under portions of the curve will correspond closely to the area of the histogram bars for that part of the horizontal axis. The total area under the curve can now be taken as representing unity, that is, the sum of the probabilities of each possible outcome. The histogram, which showed relative frequency, has been transformed into a chart showing the relative probability of each outcome. This is called a probability distribution. We have in fact turned the histogram into a *model* which we can use to describe other events in the past and future.

A model is a way of condensing the essential facts of a situation so that it can be discussed and manipulated without the expense and difficulty of working in life-size. This applies to statistical models just as much as it does to architects' or ship builders' models. Statisticians use many kinds of model. One of the simplest is the straight line drawn through a set of plots on a graph. The line is described mathematically, for instance, as $y = 2x + 1,000$. This might describe how the salaries of senior executives bear a fixed relationship to that of junior executives in manufacturing companies in Ireland. It says that senior executives are typically paid twice as much as juniors plus £1,000. If a typical junior is on £12,000, the senior will be on £25,000, and so on. Clearly this model is very over-simplified. A more sophisticated model would no doubt include such factors as relative age, academic or professional qualification and size and profitability of company. But as long as this simplification is borne in mind, such a model can be a useful and dynamic summary of the information. We can, for instance, use it to predict future salary levels.

There are several standard models (previously called 'ready-made suits') that statisticians use all the time. One special type is the 80/20 analysis that was described in Chapter 20. In this case the data falls into highly differentiated categories, thus enabling the investigators to discover 'the vital few'.

The binomial distribution

A model that is much used in quality control work using attribute rather than variable data is the binomial distribution. This is used, for instance, when a sample of production is taken from the line, and a certain number are found to be defective. The question is, what are the probabilities of randomly getting that number defective if the process is under control? Obviously if the probabilities are very low, it is more likely the process is out of control than that we have hit an unlucky sample.

Suppose we choose two items from a batch, which we expect to be 10 per cent defective. The probability of a single item picked at random being defective is 0.1. The rule about multiplying individual probabilities applies as in the sock hunt, so the probability of both items being defective is $0.1 \times 0.1 = 0.01$. The probability that one will be acceptable is $1 - 0.1 = 0.9$; therefore the chance of getting two acceptables is 0.81. This leaves the possibility of one acceptable and one defective. The danger here is to miss the fact that there are two different ways of achieving this mixed result. There are four possible outcomes, not three. The chance of the first way (defective + acceptable) is $0.1 \times 0.9 = 0.09$; the chance of the second way (acceptable + defective) is $0.9 \times 0.1 = 0.09$. This makes 0.18 in all for this possibility. The total possibilities can be summed up as follows:

Both defective $0.1 \times 0.1 = 0.01$
Both acceptable $0.9 \times 0.9 = 0.81$
Mixed result $(0.9 \times 0.1) + (0.1 \times 0.9) = 0.18$
Total of all possible outcomes $= 1.00$

If the sample is large, the working out becomes more complicated. The number of ways in which a mixed result can be achieved escalates alarmingly. However a remarkable invention called Pascal's triangle (named after the mathematical adviser to the aristocratic gamblers mentioned earlier) enables us to read off the possibilities. Thus there are only two possible outcomes from a sample of one, four from a sample of two, eight from a sample of three, sixteen from a sample of four and so on. The exact probabilities can be worked out by algebraic expansion of the formula $(q + p)^n$.

The normal distribution

As the sample grows larger, the shape of the binomial curve grows nearer and nearer to that of the best known general statistical model — the bell-shaped 'normal curve'. This was first discovered by astronomers, who called it the 'Curve of Errors'. They discovered that repeated fixings of the exact location of a star tended to produce a series of values which clustered round the true value like bullet holes round the centre of a marksman's target. Over and over again the graphs of the observed values clustered round the average (true) value in the now-famous bell-shaped curve. Later other scientists discovered that the same curve described other natural distributions. The height of any group of men or women, sets of data on head sizes, brain weights, the chest measurement of Scottish soldiers, the variability of manufactured products, the size of grapefruit, the lengths of ears of corn, all can be described by the normal distribution curve.

Quality in Practice

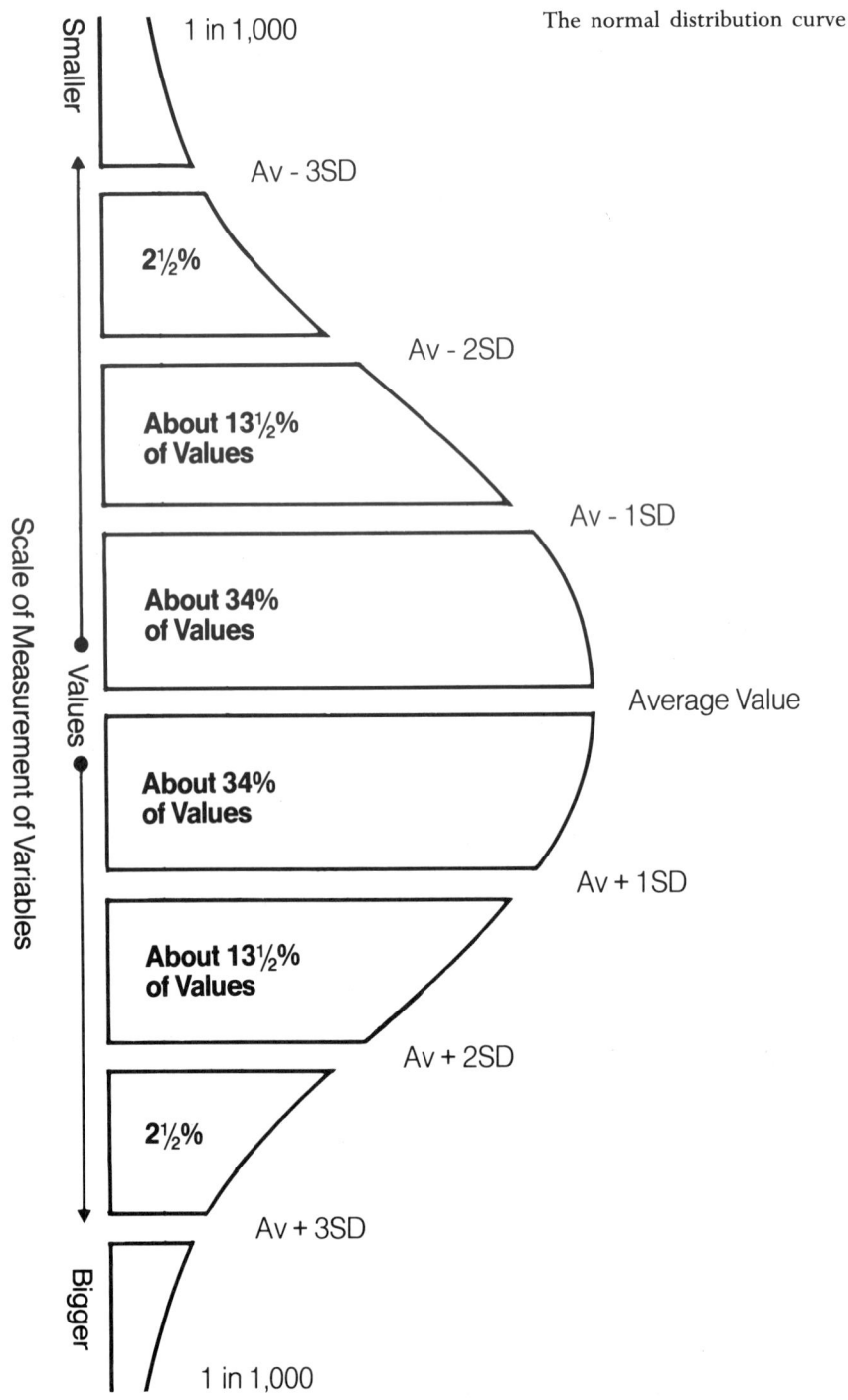

Proportions of Values Within Standing Deviation Supergroupings

The normal distribution is a mathematically defined set of relationships, as a circle is. Just as circles with different diameters differ in size, but retain the essential relationship between radius and circumference (π), so do normal curves differ. Each normal curve is defined by the arithmetic mean and the standard deviation of the underlying data. The mean is the value at the peak of the curve, which falls away gracefully and symmetrically on either side. The slope of the falling away is dictated by the standard deviation. Distributions with large means and small standard deviations will have relatively tall thin 'bells', while those with large standard deviations will have flatter curves with less pronounced peaks.

The beauty of the normal distribution is that once the mean and the standard deviation are known for a set of data that follows the model, the probabilities for a wide range of events can be worked out. Thus 68.2 per cent of the data will fall within plus or minus one standard deviation from the mean; 95.4 per cent plus or minus two standard deviations, and 99.8 per cent within three. Tables like log tables are available to give the exact proportion of the data that should fall under small fractions of a standard deviation plus or minus. These relationships are true whatever the apparent shape of the bell.

The property that makes the normal curve fundamental in statistics, however, is described as the 'central limit theorem'. This states that *whatever the underlying shape of the distribution*, a properly selected sample will enable us to make inferences and decisions about the population based only on the sample mean. The larger the sample, the nearer will the sample mean and standard deviation be to the actual or population figures. This applies for bi-modal, 'skewed', 'uniform' and all other types of distribution. This ability to make inferences from the sample without having to worry about the underlying distribution enables quality controllers to be very precise in their sampling. The standard sampling plans for acceptable quality levels are all based on this theory.

Other distributions

One property of the normal distribution is that the mean, the mode and the median are the same. This implies a perfectly symmetrical distribution. If the median and the mode are not near the mean, then the distribution is *skewed*. A typical skewed distribution would be the wages in a company, where a lot of people get relatively low wages, and a few relatively high. Another form of distribution is the *uniform* distribution, where all the possible outcomes are equally likely. A histogram recording the number of 1s, 2s, 3s, 4s, 5s and 6s from a series of throws of a dice would be (or should be) uniform.

Conclusion

Statistics is the science of handling large numbers. Like other sciences, it is unnecessary for most people to know it in detail, yet helpful to have an overview of the subject. A familiarity with some of the basic concepts is very desirable.

The first problem with an undigested set of figures is to summarise them in a representative and meaningful way. There are three ways in which this is done: the arithmetic mean (normally just referred to as 'the average'), the median, and the mode.

The next step is to recognise that two sets of figures might have the same average, yet be very different. This is the case for the sets 16, 1 and 8, 9. To learn more about our data, then, we need some measure of scatter. The simplest measure of scatter is the range, but the most commonly used is the standard deviation. This is in effect the average of the divergences of the figures from the arithmetic mean.

Another important measure of scatter, used widely in attribute-based data, is the proportion.

The measures that describe a set of data can be generalised by the introduction of the concept of probability. In a simple way this is done when someone says, 'Quality costs were 24 per cent of sales last year, so they should be more or less that this year.' The person is using the data from the past, assuming all things will be equal and applying the proportion to the future. Any set of data can be turned into a probability estimator in the same way. This is done by transforming the values achieved (e.g. the graph of the actual heights of men) into a probability scale from 0 to 1. What started as a statement that 4,036 men out of 25,000 (16 per cent) were 5 feet 7 inches becomes a statement that there is a 0.16 probability that the next man you meet will be 5 feet 7 inches.

There are a number of probability distributions that statisticians have described in detail. As far as possible the figures we have should be fitted to one of these, so that the known properties of the standard curves can be used to infer more about our own data. The best known and most widely used of these distributions is the bell-shaped normal curve. The binomial distribution, which is used for attribute data, is also widely used in quality work.

Chapter 22

Inspection and sampling

THE traditional image of the quality control function pictured the quality control inspectors as police. Their function was to examine the results of everyone else's hard work, and to consign large proportions of it to the 'outer darkness' of the rework bench. They were not as a result particularly popular members of the firm. The modern approach to quality planning concentrates on getting things 'right first time', rather than trying to inspect quality into the finished product. Starting with the quality manual, every effort is made to cut down the need for inspection. But even the extreme Japanese technique of placing the inspection responsibility on the workers themselves doesn't abolish the need for inspection techniques as such.

By inspection we mean the comparison of what has been made or purchased with what was specified. The standard of reference is the specification in the quality manual, which will also specify the techniques by which conformance or not is assessed. There are four points of inspection.

1. *Incoming materials*: A very large proportion of defects in manufactured goods has been traced to defects in bought-in materials. The control of these materials, first through supplier assessment and then through monitoring the quality of the incoming product, is vital. The inspector will be responsible for maintaining a system to prevent as far as possible products getting through to the shop floor until they have passed the specified tests. Inspection can also be carried out at the supplier's plant by inspection agencies. This can be important if the plant is far from the supplier. Most companies adopt a system of labelling for untested, approved and rejected batches. The results of the tests carried out should be carefully recorded both in the raw material file and in the vendor's file for future reference.

2. *Initial or first-off inspection*: 'First piece' inspection, as it is also called, aims to ensure that when any change in manufacturing conditions is made, such as the resetting of a process tool or the introduction of a new product, the quality will be maintained. This inspection also enables the quality team to check whether there are any errors in the design or specification, whether the right materials are being used, and whether the machine operator has correctly understood the specifications.

Quality in Practice

Testing hairdryers: note the job instruction card   *Braun*

3. *In-process inspection*: Systems for controlling and inspecting processes vary from plant to plant. They can be very complex. The purpose of this in-process inspection is to ensure that the ongoing work is kept carefully to specification. In-process inspection may be done by operators (see Chapter 5) or by inspectors alone. Inspectors may work on a machine by machine basis, taking a sample of work from the output of one machine and testing it, or they may work on a process by process system, whereby the output of one process is tested before it progresses to the next stage. When an inspector finds a machine producing defective work, the rejected items must be quarantined. There are several possible choices for unsatisfactory products at this stage, depending on the technology involved. It might be sorted, reworked, accepted 'as is' (with customer approval), ground down for re-use, or scrapped. The routes through each of these paths should be clear from the quality manual.

4. *Final inspection*: Production goods for final inspection must all have the appropriate in-process inspection approvals. The sampling plan for acceptance of the batch will be laid down in the quality manual. Final inspection will usually be responsible for ensuring that the products are accompanied by the necessary manuals and conformance certificates, and that the despatch instructions are complete.

Inspection and sampling

Measuring pipe ovality   *Wavin Pipes*

At each of these stages, the inspector is primarily looking for conformance to the specification laid down in the quality manual. This specification will be expressed in terms of a few critical qualities, such as length, thickness, weight, strength, finish, colour, taste or output speed. The specification will distinguish between various types of fault, and the action to be taken on each. A defect might be:

— *critical*

— *major*

— *minor*

— *incidental*.

Very often the difference between a major and a minor defect will depend on the spread of the defect. A few paint blemishes on an industrial tool might be minor; if the paint is very blotched, the defect will be major.

In addition to the classification of defects by degree of seriousness, there is classification of defects by type of test. The cheapest and simplest test is the test for *attributes*. This test merely ascertains whether the quality is present or not. The results are usually expressed as pass/fail or go/no go proportions, and are handled statistically by the binomial rules discussed in Chapter 21. More information can be got from a test for *variables*. This is a test that involves some

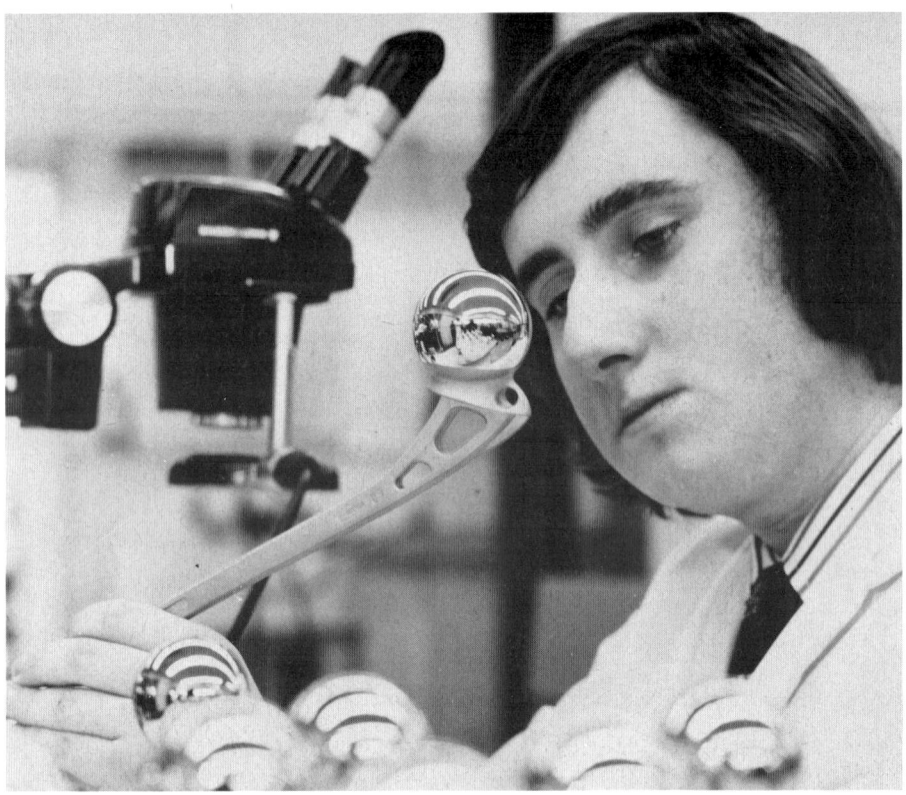

A final inspection of an endo-prothesis hip implant   Howmedica

kind of measurement, and expresses the result in figures. Typically the result would be in so many millimetres, degrees, or any other specified set of values. Variables are typically handled statistically by the normal curve and other continuous frequency distributions. The variable provides the answer to such questions as 'How long?'; the attribute provides the answer to the question 'Is it long enough?' Clearly many qualities can be tested as attributes or variables by changing the question slightly. However, there are a large number of qualities that cannot be tested as variables. Typically they are difficult to define precisely, such as the organoleptic qualities, for example taste, smell, texture.

Inspection is never foolproof. Even with the most careful and systematic operators, human inspection procedures will unearth only about 80 per cent of the actual defects. One reason for this is the way the mind and the eye relate to each other. As an experiment, read these sentences and prepare to answer questions on them.

Paris	A bird	Once
in the	in the	in a
the spring	the hand	a lifetime

Very few people indeed notice that in each of these sentences the article is repeated. The fact is that the eye and the mind working together see most easily what they expect to see. They also resist seeing what they don't expect. An experiment was made with playing cards, some of which had the colour changed, i.e. the six of clubs was red instead of black. The cards were shown quickly to the subjects of the experiment. Some unconsciously changed the anomalous cards to normal ones, either by reporting a black six of clubs, or a red six of hearts. Others saw the card as the six of clubs, but in purple or brown. Furthermore, it took longer for people to name the abnormal cards than normal ones.

The technique of seeing is clearly more complicated than it would appear. It seems that the mind carries a set of images, which are then matched against the perceived object. If no image matches, the mind tends to opt for the 'best fit'. It appears to resist new visual experiences. Another experiment, frequently conducted by Dr J. M. Juran, one of the leading American quality control experts, asks the subjects to examine the following sentence:

FEDERAL FUSES ARE THE RESULT OF YEARS
OF SCIENTIFIC STUDY COMBINED WITH
THE EXPERIENCE OF YEARS

and count the number of times the letter F appears. In group after group, the average result is that only about 80 per cent of the Fs are found. (There are six altogether. As an illustration of the importance of inspection technique, it is more effective to search for the letter from the end of the sentence backwards, thus minimising the distraction of the meaning.) Research into the source of inspection errors has divided them into three categories.

1. *Technique errors*: Lack of knowledge of test techniques, lack of physical aptitude (such as short sight), or lack of skill in the operation of test instruments can all be remedied by training programmes.

2. *Inadvertent errors*: Even the most skilful inspectors make these errors unawares. The precise psychological mechanism that causes these errors is as yet unclear. There seems to be no way that they can be quite eliminated, though well designed test systems and routines can go some way towards that goal.

3. *Conscious errors*: These errors are usually caused by the pressure of the firm's operating systems on the inspector's judgement. Under pressure from the production manager, defective products may be accepted; the inspector may 'flinch', by reading slightly over-tolerance products as within tolerance, thus causing an anomalous build-up of sample values just inside the tolerance level; the inspector may take short cuts by ignoring certain procedures, or speed up the inspection process by taking all the sample for a batch from one package.

Quality in Practice

Human and mechanical operations are variable and error-prone. That is unavoidable. Inspection is intended to control that error. Unfortunately, as we have seen, 100 per cent inspection is rarely 100 per cent effective. For human and operational reasons defective products will slip through. With 100 per cent testing there is no way of assessing the extent of this human fallibility. Another disadvantage of 100 per cent testing is that because it imposes such a burden of work, it can prevent really extensive tests being carried out. Finally, in some cases, such as foods and bullets, the only 100 per cent way to test the product would leave nothing to sell. For all of these reasons most quality control programmes are based on sample testing and inspection schemes rather than 100 per cent testing.

Acceptance sampling

The objective of acceptance sampling is to determine whether a batch can be accepted or not. The decision is taken by inspecting a sample drawn from the batch. Sampling is an everyday activity. Cooks taste soup to check its flavour; buyers of fruit check ripeness before buying; doctors take blood tests; and readers check a few pages before taking a book out of the library. The first problem in sampling is to ensure that the sample correctly represents the *population*, which is the statisticians' word for the large group from which the sample is drawn.

A sample can be chosen in three general ways.

1. *A convenience sample* is where the units are selected because they are easily got, rather than any more scientific reason. This is not necessarily a bad thing; this kind of sample is almost certainly cheaper than any other method, and may well be quite satisfactory, but is very likely to contain hidden bias.

2. *The judgement sample*, by which elements are picked according to the picker's expert knowledge of the population, is more reliable. This may be appropriate where the diverse elements in the population are difficult to quantify or group. This is common in industrial environments, where the effort of elaborately structuring the population would hardly be repaid. In the typical case, the quality personnel simply try to pick elements in the sample from as wide a range of cartons, boxes or other sources as possible. The sample is, however, to some extent an extension of the picker's view of the population, and will be biased accordingly.

3. *Random samples* are the most satisfactory of all. In this case care is taken to ensure that each unit in the population has an equal chance of being selected. In order to ensure this, the elements in the population have to be ordered in some way. This might be by strata, by group or by unit. The sample is then selected, perhaps using random number tables, from each set. Strictly speaking, a sample

TABLE I — Sample size code letters

Lot or batch size			Special inspection levels				General inspection levels		
			S-1	S-2	S-3	S-4	I	II	III
2	to	8	A	A	A	A	A	A	B
9	to	15	A	A	A	A	A	B	C
16	to	25	A	A	B	B	B	C	D
26	to	50	A	B	B	C	C	D	E
51	to	90	B	B	C	C	C	E	F
91	to	150	B	B	C	D	D	F	G
151	to	280	B	C	D	E	E	G	H
281	to	500	B	C	D	E	F	H	J
501	to	1200	C	C	E	F	G	J	K
1201	to	3200	C	D	E	G	H	K	L
3201	to	10000	C	D	F	G	J	L	M
10001	to	35000	C	D	F	H	K	M	N
35001	to	150000	D	E	G	J	L	N	P
150001	to	500000	D	E	G	J	M	P	Q
500001	and	over	D	E	H	K	N	Q	R

Quality in Practice

TABLE II — Single sampling plans for normal inspection (Master table)

(US MIL-STD 105D)

A sampling plan for attributes: products classed as defective or non-defective

⇩ = Use first sampling plan below arrow. If sample size equals, or exceeds, lot or batch size, do 100 per cent inspection
⇧ = Use first sampling plan above arrow.
Ac = Acceptance number
Re = Rejection number.

drawn 'at random' from the population, without the control of random number tables and other objective randomising devices, is not truly random.

A truly random sample has the best chance of representing the population, because it is free from *sampling bias*. Nothing in the actual selection of the sample tends to bias the data. Unfortunately, however, it is possible that a random sample may suffer from *sampling error*. This occurs when a scientifically selected sample fails to represent the population by chance. Bad luck has simply presented us with an unrepresentative sample. Just as in the coin-tossing example discussed in the previous chapter, every possible sequence of results has a small but equal chance of being selected at random. The larger the sample, the less likely it is that sampling error will occur.

The first requirement in creating a sampling plan is to decide the *acceptable quality level* (AQL). This is the maximum percentage of defectives that can be considered satisfactory. It might be 0.15 per cent for major defects and 10 per cent for minor defects. The AQL will often be set by the customer.

The next stage is to consult the sampling tables specified in the various standards. The first table gives the code for the sample to be picked from various batch sizes, depending on the AQL and the level of inspection. Thus for a batch size of 10,000, an AQL of 1.0 per cent and General inspection level II, the second table specifies a sample of 200. If six or more defectives are found, the whole batch should be rejected.

A more sophisticated version of the single sampling scheme described in IS 301 is the multiple sampling plan. In this scheme there are three possible decisions after examination of the sample: accept, reject or take a further sample. If a further sample is taken, the results are added to those of the first sample and an accept/reject decision is taken. The advantage of the multiple sampling scheme is that inspectors are occasionally reluctant to reject entire batches that only just fail. The opportunity to re-examine the data accords with human psychology. There are also potential cost and time savings, since the first sample is usually about two-thirds the size of an equivalent from a single sampling plan.

Another way of controlling the cost of inspection is to regulate the level of inspection. IS 301 specifies three levels — reduced, normal and tightened. The less the degree of inspection, the smaller the sample. This of course increases the risks of accepting substandard batches. Changes from one level to another are usually triggered by a record of continuing good or bad quality performance. The Standard suggests that tightened inspection should come into effect when two out of five consecutive batches have been rejected on normal level. When five consecutive batches have been accepted, inspection should move back to normal. Reduced inspection should be introduced when production is at a steady rate, when at least ten consecutive batches have been acceptable, and the total number of defectives is less than certain specified quantities. As soon as a batch is rejected, or production becomes irregular, normal levels should be resumed.

Chapter 23

Documentation and control charts

DOCUMENTATION is the lifeblood of quality control. The whole strategy of moving away from the 'Ah, sure, it'll do' approach depends on a clear system of standards and minimum achievement levels, and then on a set of records that monitors these standards. Companies also have external pressures on them to maintain rigorous documentation standards. As more and more government and commercial organisations adopt supplier assessment schemes, companies selling to those sectors have to show that they have carried out desired quality controls. To a lesser extent, consumer organisations are demanding more documentation as a means of providing more information to consumers. Another motivation, particularly in the food and drug industries, is the weight of government regulations demanding strict quality standards. And finally the new law on product liability makes it highly desirable for manufacturers to keep defensive records of their systems.

Documents can be divided into three groups. These are:
— *System descriptions*: Quality manuals, workflow charts, standard operating procedures, sampling plans, etc.
— *Process documents*: Batch records, process control charts and labels which are actually part of the production process.
— *Summary records and plans*: Supplier assessment documents, training schedules and records, calibration records, etc.

System descriptions

The fundamental system description is the quality manual, which is dealt with in Chapter 16. This lays down both the policy and practice of quality management. The size of the manual can vary from a modest system description of a not very complex process to the elaborate multi-volume GMP manuals of standard operating procedures used by some drug firms. The more extensive the activities of the firm, the greater the need for a statement of quality regulations. But as well as being a bank of knowledge about production practices, the quality manual is also summary of standards and aspirations. With its associated workflow diagrams and staff responsibility charts, the quality manual is the basis of the entire structure of documentation.

One type of system description that is sometimes forgotten is the signposting of areas in the plant. Accurate and frequently updated labels and signposts make it clear to the workforce and to visitors that a planned quality system is in force. The distinction between, for instance, incoming materials areas that contain passed and not-yet-passed stores, quarantine areas for rejected goods, clear labelling of sterile areas or areas of limited access all make up an important part of the system description.

Process documents

The quality manual will specify the use of various process documents. These are the records that accompany the product through the plant, or record its passing. The simplest form of process document is the label. This is usually stuck directly on to the carton or shipper in which the product is stored.

The next type of process document is the batch or lot record. One typical form records the date and time of manufacture of the batch, the operators and machines involved, the raw material batches used, and the results of any tests carried out by the operator. The quality staff record the results of any tests they carried out. This record will include the sampling plan used, and the kinds of defects found, classified as critical, major and minor and by type. The quality manager is often required to sign the batch record to confirm that the department has approved the batch. The batch number is usually recorded on the label so that products waiting in the warehouse for onward shipment can be monitored.

Exhibit: Traceability at Chemoflon GmbH, Tallaght, Co. Dublin
All goods sold by Chemoflon have a 'QC Checked' label attached to each item, during final inspection, when they meet the required standard. The label has a six digit code printed on it. The digits represent the following:

First two digits	– the inspector
Third digit	– the year in the decade
Fourth and fifth digits	– the week number in the year
Sixth digit	– the day of the week

In the situation where a product is found to be defective by the customer, the siting of the QC Checked code will allow the quality controller to trace the history of that product through the complete manufacturing system to receipt of raw material.

The QC Checked date signifies a particular Final Inspection report. The Inspection Report will show the QC number and the Job number for that product. From these the date and the finishing treatment received by the product can be found, along with the various production machines on which the job was produced, together with the names of the operators, process inspection reports at various stages, chemical batch numbers, yarn lot control numbers, incoming material reports, date of receipt and suppliers of raw materials.

An important facet of the traceability system is that each person who makes a check, whether they be operator, inspector or supervisor, signs their name or initials to give assurance that the check was effected. In this manner each person becomes accountable to the next operation and will not let a

Quality in Practice

R. & A. Bailey & Co. Ltd
Out of Specification Report

Number _____

Date _____

This form to be used for raw materials or bulk product prior to bottling.

MATERIAL _____

QUANTITY _____ Ref _____

SUPPLIER _____

FAULT(S) _____

**POSSIBLE/
PROBABLE
CAUSE** _____
_____

RECOMMENDATION _____
_____
_____

Signed _____

Dated _____

Informed _____

Instruction _____
_____

Signed _____
Technical Manager/Quality Manager

Circulation:

Packaging Materials (Naas Road/Nangor Road)
M.O'L. P.O'R. E.H. N.K. J.G. P.G. M.S. F.C.
Product Ingredient/Bulk Product
E.H. N.K. P.G. D.O'C. A.L. F.C. B.O'H. U.G.

Out of Specification Report is a control mechanism and source of information.

R. & A. Bailey

product through unless they are truly satisfied with the quality standard. This accountability leads to a greater sense of awareness and pride in the work each operator produces.
Source: Denis Keogh, Chief Executive, Chemoflon GmbH, Co. Dublin

The batch record is critical for the achievement of a key objective in quality systems, traceability. This is the way in which the quality system can effect feedback. Only if complete and accurate batch records are kept can the firm learn by its mistakes. If, for instance, a customer complains of poor quality, it should be possible to identify every input into that product. Thus the raw materials used (and of course the vendors who supplied them), the operators and machines, the quality staff who tested the product, the tests used, and the packing and despatch routes should all be easily recoverable. The complaint can then be used as a stimulant to re-examine the system to see whether possible sources of defects can be controlled. With clear records, the faults in the system can be diagnosed. For instance, a complaint may expose the fact that the incoming materials assessment wasn't sufficiently rigorous, or that the wrong tests were specified.

While batch records follow the fates of individual lots, process control charts monitor the variability of machines. The assumption behind control charts is that there are two sets of causes of variation. The first is random variation, caused by many tiny shifts and tolerances which are part of the machine's design. They cannot usually be economically eradicated. If only random causes are present, the system is operating as well as it can. If the product is still not within specification, the system must be changed. The second set of causes of variation contains the assignable causes. These are the ones about which something can be done. They are typically single factors, such as wrong machine setting, operator inattention, or wrong raw materials.

All machine processes tend to drift out of specification if left unattended. One object of process control charts is to display the drift, so that the operation is brought back on course *before* the batch is complete and rejected on inspection. The design of control charts should enable operators to distinguish between random and incurable variations in the system's output and movements that signify a change in manufacturing conditions. Modified control charts can also be used to monitor non-manufacturing processes. For instance, the percentage of defective deliveries by vendor can be charted, or the speed of turnround of response to customer complaints, or the number of hours lost per week through down-time. However, it should be noted that these events probably do not follow the normal distribution model, so the setting of the control limits (described below) would have to be done differently.

There are many types of control charts, but they are all set up by the following basic procedure.

1. *Measurement*: Decide what is to be measured and how. This might be a variable, such as length, diameter or temperature, which will be measured in figures. Or

Quality in Practice

Checking the specification during a complex assembly job   *Measurex*

it might be a proportion of the sample possessing an attribute, such as a specified colour, a specified density or any yes/no property, including such attributes as being delivered on schedule. Attribute charts, however, provide less information than variable ones. When a problem arises, they very often have to be supplemented by variable data to enable exact causes to be discovered. The first priority should be given to setting up controls for characteristics which are currently causing trouble.

2. *Choose the statistic*: Depending on the process to be controlled, decide on the statistic to be charted. This might be the values derived from one individual picked from the line, from the average of a sample, from the range, the proportion defective, the total number of defects per fixed sample, or many other possibilities. At the same time a decision should be made as to how often the sample value should be taken. This of course will depend on the changeability of the system. A stable system may need to be monitored only once or twice a day, whereas a volatile one may need hourly checking.

3. *Begin the chart*: Draw the first elements of the control chart. The horizontal line at the bottom records the observation number or time. The vertical line on the left-hand side gives the scale, and the middle horizontal line is the central line, representing the target value. This is the line around which the operator expects values to cluster. Of course, if the machines are set correctly, this value should also be the desired process average, or specified value.

4. *Plot the first few values*: The random variability inherent in the system will quickly become clear as the plotted values fall above and below the central line. For

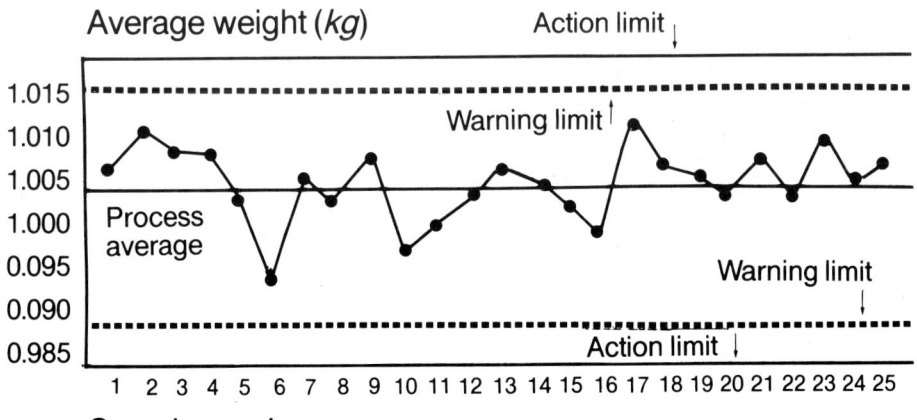

A standard control chart for average

short runs, pseudo-trends may occur, as subsequent values appear to be moving in one direction. If this is just a random happening, the 'trend' will quickly reverse itself, and the values continue to fluctuate around the central line. However, the trend may represent real drifting in the output. The problem is to distinguish assignable from random movements.

5. *Set the control lines*: The solution is to set control limit lines parallel to the central line. If the plotted values approach one of these limits, the system is likely to be drifting away from specification and should be adjusted accordingly. The lines are set according to the assumptions of statistical probability discussed in Chapter 21. The random variability of the machine process is assumed to follow the bell-shaped normal distribution. The further away a reading is from the process average, the less likely it is that it would occur at random. The formula for the normal curve suggests that 99.7 per cent of random fluctuations will occur inside the area covered by plus or minus three standard deviations. A reading that breaks a limit line calculated at three standard deviations of the statistic used has therefore only a 0.3 per cent chance of occurring at random. Limit lines should normally be set by a statistician. It is almost certainly therefore a real change in the production process and should be investigated.

The basic structure of control charts can be developed in many ways. If the cost of changing or investigating an apparent shift in production output is not high, it might be better to set the control limits at plus or minus two standard deviations. This would be a more sensitive setting. Like a wind gauge set on the top of a hill rather than in a valley, it would highlight more random and irrelevant movements, but also detect real changes more quickly. Control limits set at two standard deviations plus or minus mean that 68.2 per cent of random movements are contained inside the control area, and 31.8 per cent outside. In approximately one-third of cases therefore apparent change-of-state shifts would be due to random factors. Some charts use the two standard deviation level as a warning limit, with the three standard deviation level as an absolute control limit.

The designer of control charts has to balance the cost of investigating mere random movements that are part of the system with the risk of producing non-conforming product.

Types of control chart

1. *Control chart for averages*: This is commonly known as the $\bar{x}$ chart, from the character $\bar{x}$ which is the usual sign for an average. $\bar{x}$ is read x-bar. This is the most usual form of control chart. Small samples of two to five items are taken at regular intervals from the line and examined for the critical characteristic. The average (mean) reading is calculated and plotted on the chart. There is a play-off between the size of sample, the frequency of sampling and the sensitivity of the chart. The smaller the sample, the larger the standard deviation will be,

and so the wider and therefore less sensitive the control limits. If the test method makes it impractical to test large samples, the same effect can be got by increasing the test frequency.

2. *Control charts for range* (also called *R* charts): When the spread of values is more important than the central value, the chart is used to plot the range of the sample. This is usually represented as a bar connecting the two extreme values. Very often this is combined with the average, so the bar shows extreme range values and, marked in the middle, the average value. If the sample is large enough, it might be sensible to use the semi-interquartile range rather than the full range, thus avoiding extreme readings.

3. *Control chart for standard deviation*: Often used with the control chart for averages, the standard deviation chart provides a better measure of the deviation inside the sample than the range.

4. *Control chart for individuals*: In some cases it may be desirable simply to take one item off the production line at frequent intervals and plot the values derived from that. This chart, as explained above, will be subject to relatively violent movements, since it lacks the damping effect of the averaging process. A strong advantage of this kind of chart is that it avoids confusion between specifications and control limits. For $\bar{x}$ charts, there is an assumption that the actual real average may be greater or less than the sample average. This probability follows the normal curve. Thus if the average is near the specification tolerance, it is very possible that the real (population) average may be beyond the limit.

5. *Control chart for fraction defective* (also called the *p* chart): If the output is assessed simply on a go/no go basis, the proportion of the batch found defective can be charted. This form of chart is most effective when the samples are large, that is more than fifty at a time. Clearly if the sample is small, and the proportion of actual defectives is also small, there is an increased likelihood of a sample showing no defectives and therefore misleading the quality staff. If the samples are all the same size, the number defective can be plotted instead of the proportion. This is called an *np* chart.

6. *Control chart for number of defects*: If it is possible for a product to be defective in various ways, or to have minor or incidental defects short of scrapping, the total number of defects may be charted. Statistically it is important that the defects be caused independently. If several defects are likely to arise from one cause, the chart becomes unusable. This chart can be designed to record the total number of defects per a fixed sample (a *c* chart), or the number of defects per unit if the sample varies (a *u* chart).

7. *Cumulative sum control charts*: Just as the average is more sensitive as an indicator of trend than the individual value, because it encompasses more data, the cumulative sum chart is more sensitive still. This chart is calculated by taking the movements of the sample value from the average. This will result in a positive or negative figure, which is added to the previous reading. If the process average is 10, readings of 9, 9, 8, 11, 12, 13 will result in the chart being marked cumulatively $-1, -2, -4, -3, -1, +2$. The effect is to reinforce the impression of trend. In this kind of chart, the usual control limit lines do not apply. This is because the slope of the data line is critical, not the actual position. The steeper the slope, the more significant the trend. *Cu-sum* charts, as they are sometimes called, are often read with a transparent overlay or mask which indicates when the slope has reached statistically significant levels.

Summary records and plans

As well as records that are part of the production process, the quality management system needs summary records. These are the records and files kept by the quality staff to help them carry out the specifications in the quality manual. One role of summary records is to enable quality staff to monitor their own activities. Thus the manual may lay down a supplier assessment system with an audit schedule, and perhaps some kind of control chart monitoring performance. The actual results of the supplier assessment as carried out on a specific supplier would be stored in the summary records. The full set of these provides back-up and justification to the approved supplier list.

Equally the quality manual may specify that every operative be given certain types of training. The summary records will specify exactly what training Sean, Anne or Colm have received, and what is planned for them. The records of the progress and results of the calibration programme fall into this category as well.

Another type of summary record is the quality performance summary. This is typically provided to general management on a weekly or monthly basis. It summarises the quality performance for the previous period in a series of accepted quality indicators, such as defectives per thousand, proportion of defectives, etc. Results of special quality surveys, audits and investigations should be part of this report. Estimations of quality costs should also be included.

Chapter 24

Computers

WHEN the first computers were developed in the 1940s they were extremely cumbersome and expensive. They broke down almost daily, and were difficult to program. So much so that the British government of the day declined to invest in this new technology because they were advised that 'there couldn't conceivably be a need for more than six computers in the whole country'.

Things have moved on a bit since then. Nowadays most offices have access to computer power, even if it is only used for word processing and accounts. Computers can process data, complete complex calculations rapidly, tabulate and graph information, and provide instant reports. Properly used, they can increase the effectiveness of the quality function immensely. Unfortunately most computers are very much underused, or used quite inefficiently. It is therefore necessary to understand exactly what they can do and, just as important, what they can't.

The structure of computers

Computers are divided into three types: mainframes, minis and micros. The larger the computer, the greater the cost and the greater the power. A mainframe usually has the capacity to process all the accounts and other work of a large company. These systems are typically used for airline booking and banking applications. It can handle, apparently simultaneously, as many as 200 or more users at once. This market is dominated by IBM machines. Minis are smaller and less powerful than mainframes, but they still cost more than £20,000. The micro has taken over many of the tasks previously and wastefully done by larger computers. This kind of machine has also greatly extended the accessibility of computing. On the other hand, a micro is sometimes too small to handle all of one person's work, and very often cannot print out data and calculate at the same time.

The principles on which computers work are universal. Whatever the size differences, and the other differences derived from kinds of use, the basic structure or 'architecture' of all digital computers is the same. The only thing any computer can do is handle information in very small amounts very fast. Internally the machine works in binary values, represented by the presence or absence of a voltage on a circuit at any time. Eight of these binary digits ('bits') make up a byte, which

Quality in Practice

represents one alphabetic character. The word 'quality' thus requires eight bytes of storage. Information expressed in this binary form is referred to as 'digital', or 'digitalised'. The computer manipulates these bytes of data, one or more bytes at a time. If this was done at human speed the system of working would be impossibly clumsy, but since computers work at speeds of 5 million operations a second or more, the clumsiness is more than compensated for by the immense speed.

The working of the computer is controlled by software. This is a set of instructions called a 'program'. These tell the machine, step by step, what to do. Without software, the computer is as inert as a TV without electricity. Furthermore, the software has to be suitable for the hardware, or the physical machine, of which each design is slightly but significantly different.

A computer has three basic hardware elements. These are:

1. *The processor*: This is the element of the computer which does all the work. It is also called the Central Processing Unit or CPU. The components of the CPU are the Arithmetic/Logic Unit (ALU), the registers and the control circuits. The ALU is the part of the computer where the calculations and logical decisions are actually carried out. It is made up of a number of circuits consisting of logic gates through which bits of data pass and by which they are transformed. In fact the ALU can perform only a very limited number of operations. It can add

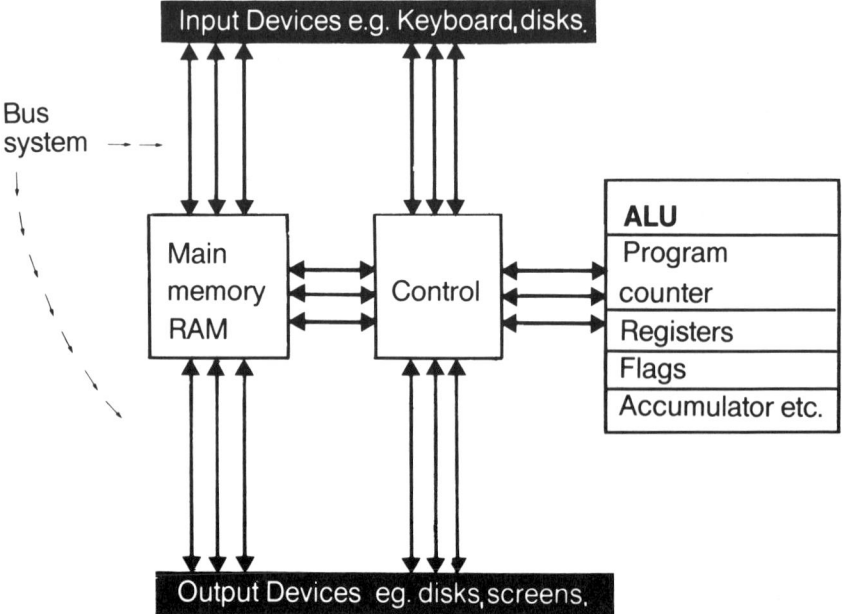

Standard computer architecture

two numbers in binary; it can subtract two numbers, which it does by making one of them negative and then adding. It multiplies and divides by adding or subtracting over and over again. It can compare two sets of 0s and 1s to see whether they are the same. Everything the computer program does has to be built up of combinations of these functions. What is more, every action has to be processed one at a time through the ALU. This bottleneck is acceptable only because of the speed of processing.

The registers are temporary stores for information, instructions and memory addresses as data is moved around the CPU and to and from the memory. One register, called the accumulator, stores data that the ALU is working on. Another keeps a count of the line of the program currently in action. The instruction register takes the actual instruction represented by the program line number and presents it for action to the ALU. Other registers store information as to whether the figure stored in the accumulator is negative or positive.

The final element in the CPU is the control function. The basis of the control is a clock. This sends a regular stream of high speed electronic pulses to the control circuits. The computer on which this book was written works at 4.77 million pulses per second. On receipt of the pulse, the control circuits move everything on one step. The next instruction is obtained from main memory and decoded so that the logic circuits of the ALU can handle it; data is recovered from the main memory for the instruction to operate on; output signals are despatched to the rest of the computer. Thus the whole machine works in series of extremely rapid jerks, with sets of data being quickly moved round the circuits.

2. *The bus system*: The bus system is the name given to the set of parallel connections that allows data to move between the memory and the CPU and inside the CPU itself. There are three types of bus, which always run together.

The control bus carries the clock impulse signals from the control element of the CPU to parts of the computer. The data bus carries groups of bits of information to and fro. The address bus runs with the data bus to inform the machine where a piece of data is to go. The size of the address bus dictates the ultimate size of the main memory of the computer. A sixteen line address bus can separately identify 65,536 individual locations (or 64K, since $1K = 2^{10} = 1,024$ bytes and therefore $2^{16} = 65,536$ bytes). A 19 line bus can address 2^{19} locations, or 512K.

3. *The main memory*: The computer's CPU handles data in very small parcels. Once one operation has been completed, the CPU will ask, 'What now?' and the next instruction has to be fetched from memory. The results of the previous instruction may be stored in memory at the same time. The computer's memory is made up of two types, read-only memory (ROM) and random access memory (RAM). Read-only memory contains fixed and non-erasable instructions telling the computer how its internal functions work. This includes such things as how

Quality in Practice

information is displayed on the screen, and how data is accepted from the keyboard. Random access memory is the computer's scratchpad. The current program is stored there, as is all the incidental data being worked on. The CPU can handle data only in very small quantities, perhaps only one letter at a time. Each new letter has to be separately fetched from the main memory of the computer, where it is stored in a certain address. The program will specify 'take data stored in memory location 8192 and bring into CPU for processing'. The data is then read from the specified location and brought along the bus system to the CPU.

Input/output and back-up memory

The second set of hardware that the computer needs is that relating to input/output and secondary storage. Programs and data have to be put into the machine, and the results of the computer's work expressed in humanly usable form. Typically the primary input device is a keyboard, and the input data is simply typed in. A visual display unit (VDU) screen, like a specialised television set, shows what is being typed, and is often used as an output device as well. Other kinds of output device include printers, plotters, actuators to throw switches, etc. and special devices (modems) for sending digitalised data down telephone wires to other computers.

Using computers to test memory circuits   *Wang*

The size of main memory directly addressed by the bus system is limited by the number of lines in the address bus. A standard micro such as the Wang PC has a twenty line bus, which allows it to address 1,024K bytes of RAM. This space is eaten into by the program, at least some of which has to be loaded into RAM before operations can start. RAM is also volatile, in that it does not continue to store data once the computer is switched off. It is therefore necessary to have secondary memory to back up the main store. These back-up devices enable computers to preserve and allow access to historic data.

There are various secondary memory devices, of which the most common is disk storage. Disks (also called diskettes and floppy disks) are made of a thin film of magnetic oxide on a base of thin plastic. They are sheathed in a protective case. They might be 8 inches, 5¼ inches or 3½ inches in size depending on the machine they are used in. They are inserted in the disk drive unit, and circular tracks of data are stored on them by sensitising (or not) bits of the magnetic oxide. The disk rotates under a movable read/write head, so that data is very quickly accessed. The most common disk at present is the 5¼ inch type, which can hold up to 360K bytes of data. This is about 50,000 words, or two-thirds of the length of this book.

Data is stored on the disk in concentric tracks, which are themselves made up of sectors. One widely used disk operating system lays down nine sectors to a track. When data is 'saved' from the main memory to disk, it is not necessarily laid down into the tracks in a predictable order. Sectors and tracks are filled up as may be, and a record kept of the order in the directory, which usually takes up the outside track. If data is to be deleted from a disk, the entry in the directory is simply erased. When new data is being saved, the lack of an inhibiting entry in the directory means that the new data is allowed to over-write the old. The way data is laid down on to disks is by no means common from one computer to another, although machines that share a *disk operating system* such as MS/PC-DOS should have no problems with compatibility.

If more storage is needed than can be provided on a floppy disk, then hard disks are required. These operate in the same way as floppy disks, but they are fixed into their case. The read/write head skims a micrometre over the top of the disk. This enables the data to be packed in more tightly. Hard disk systems can as a result store very large amounts of data. There are many other storage methods either used or being developed, including magnetic tape, read-only compact disks, and write-once laser disks.

Programs

A program is a sequence of instructions that is fed into the computer at the start of an operation, and then governs the flow of activity of the hardware. There are various types of program. The computer's ROM will certainly contain various elements of 'intimate software', which are intrinsic to the machine and provide the basic operating routines. An example of this is the so-called 'bootstrapping routine'. This is a short built-in program that gets over the paradoxical difficulty that a computer can do nothing without a program, not even read a fed-in program. Other activities controlled by the monitor, or operating system ROM, include loading sections of memory from secondary memory storage, interpreting commands from the keyboard, and controlling communication with input and output devices.

The monitor is special to a particular machine or design. The next stage of generality is the operating system, which consists of a number of mini-programs covering functions that computer users regularly need. The existence of operating systems prevents programmers from having to write special programs every time they want for instance to store data on to a disk, or copy a file from one part of a disk to another. This naturally speeds up the writing of special task programs. Another function of a typical operating system is supporting (i.e. translating into the computer's own code) generalised programming languages such as BASIC, Fortran, Pascal, etc. Because the operating system deals largely with the way the special program relates to the physical functions of the computer, such as data file handling, it enables programmers to write special programs that are usable by several machines, without having to know about the inner details of each. A feature of growing importance for micro users is the kind of operating system that allows different machines on the same network to talk to each other.

For the ordinary user, by far the most important software category is that of applications programs. These are the programs that help the user to perform work (or play games). Programs fall into two categories. Generally available programs perform functions such as word processing, accounts and financial planning, database and information management, and statistics. The second category is software written with a particular industry in mind. These range from integrated systems for auctioneers and architects to programs for travel agents and surveyors.

Exhibit: Typical quality control programs
Q-Cost IV
This program produces a quality cost analysis report on a monthly or twelve monthly basis for up to forty individual quality cost accounts with Pareto and trend analysis summary reports against sales, direct labour, indirect labour, conformance and non-conformance. Menu driven and printout or reports at each stage if required.
Machine: IBM PC, XT, AT and compatibles
Operating system: PC-DOS
Memory: 64K
Storage: 5¼ inch disk
Price: Stg £350

Gauge and Calibration Control
Allows the user to keep an unlimited number of gauges or equipment on file with the following

information for each: gauge name or serial number; customer name; associated part number; drawing number; drawing revision number; date of calibration; next calibration due; calibration interval; calibration procedure/remarks. Menu driven, printouts if required.
Machine: IBM PC, XT, AT and compatibles
Operating system: PC-DOS
Memory: 64K
Storage: 5¼ inch disk
Price: Stg £350

Quality Alert
Statistical quality control suite of programs for: variable data control charts including x-bar; standard deviation; moving average; moving range; attribute data control chart; p chart; c chart; u chart; process capability for variable data.
Machine: IBM PC, XT, AT and compatibles
Operating system: PC-DOS
Memory: 64K or more
Storage: 5¼ inch disk
Price: Stg £850

Source: *PC Buyers Guide* 1985

In general, there are two aspects to the use of computers in quality control. Computers can be used to improve the quality control function itself, and they can be used to provide a better quality product, by computer-aided design and computer-aided manufacture (CAD/CAM).

Computers in quality assurance work

Computers are used in seven main areas of quality control work. These are:
— data collection
— data reduction, analysis and reporting
— real-time process control
— automatic testing and inspection
— statistical analysis
— information retrieval
— quality control of production.

The best use of computers in quality control work undoubtedly comes from environments where the basic production control is also on computer. Unfortunately recent studies suggest that less than 50 per cent of companies use computers for production planning and control, and fewer than 10 per cent of companies employing less than 150 people allow supervisors and first level management to use computers. In quality assurance work, the following uses can be identified for computers.

Quality in Practice

Inputting test data to an IBM PC *Semperit*

1. *Data collection*: Computers are well suited for the collection and storage of data, because they are quick, quiet, compact, and easy to refer to. Transmission of the collected data is simple. A major plus for a computerised system is that information need be input only once. Once it is in the computer it can be used for analysis, for storage, and as part of the regular information output.

2. *Data analysis*: The computer can analyse data in different formats, and readily rework analyses if necessary. All regular reports and plots can be performed automatically. In the area of complaint analysis, for instance, the computer can rapidly connect the complaint with the batch records, with the suppliers for that particular batch, with the problems previously experienced with that batch, that supplier or that distribution chain. All the usual supplier ratings can be automatically updated on receipt of the batch, so that the very latest information is available to the purchasing and inspection people. The speed with which it works makes it possible to undertake relatively elaborate analysis. A major advantage of computer systems is that they can be programmed to report on an exception-only basis, so that only non-conforming events are highlighted. The flexibility of the system encourages reworking and re-analysis of data.

3. *Process control*: Automatic systems which measure, analyse and control process variables have been developed for products such as beer (as in the Murphy Brewery in Cork) and paper-making. When correctly set up, the computer can actually achieve genuine 100 per cent inspection, which is humanly impossible. Computers are particularly useful where interconnected variables are being measured, where time delays make human measuring too slow for the process. Another use of the computer in process control is to integrate the production of labels and batch records with testing and process control. The label can be economically made to carry more information in this way. The same information can be stored in the batch file without further effort.

4. *Automatic testing and inspection*: Robots and other sophisticated systems are expensive, but they can provide the unvarying 100 per cent testing facility that may be required. This is particularly useful in dangerous or highly unpleasant circumstances. Computer-controlled test and inspection equipment should result in lower operating costs, improved test accuracy, and automatic calibration.

5. *Statistical analysis*: Regular and special reports can be quickly produced for all areas, such as scrap, reject and rework percentages; quality control charts; supplier ratings; testing results and quality costs. More general information summaries, such as the Quality Index used by System Industries in Co. Dublin, are limited only by time and imagination.

Exhibit: A quality index
The quality index is calculated by categorising all defects found at the final quality assurance inspection on new product, using the defect categorisation chart. The formula is as follows:
Quality Index = $100 (1 - (1I + 0.2m + M + 2C)/N)$
where
I = number of incidental defects
m = number of minor defects
M = number of major defects
C = number of critical defects
N = number of items shipped.
This gives a weighted index of quality at final quality assurance inspection, with the volume of product shipped as the denominator. If the index falls below 95, we have a serious warning of deteriorating quality. If it falls below 90, we have a serious problem to be corrected.
Source: Mike J. A. Neary, Plant Manager, System Industries Europe

Computers in production

One of the most important trends in modern production is the increasing use of computers in the design and manufacture of products. The various techniques differ in their impact on the quality control function.

Quality in Practice

Use of a robot handling device (on left) allows the operator time to control quality   *Nypro*

Humans generally do not perform well in tasks that are time-consuming, error-prone or boring; they do perform relatively challenging and innovating tasks well. Programmed computers and robots are exactly the opposite. Robots are increasingly used to perform relatively straightforward production tasks. As long as the environment and the inputs remain identical, a computer or a robot will repeat the output exactly (with allowances for the random variations of the system). Computers are also increasingly being used for various monitoring tasks. Direct digital control is where the computer actually controls the production process. A sophistication of that is supervisory computer control, where the computer collects data from the production process and controls it not only according to physical parameters, but taking into account performance optimising factors relating to profit and loss. Computer process monitoring is more a control function than a production function and as such has a direct input into the quality assurance area.

The inspection task changes subtly in respect of these systems. If the process can be trusted to run correctly once it is correctly set, the first inspection becomes critical. If the first products are satisfactory, then a relatively low level of statistical process control can be maintained. This is quite different to the controls required for a very human labour intensive operation, where high levels of inspection need to be maintained continually.

PART IV

The quality environment

Chapter 25

The economics of quality

THE purpose of quality planning is to produce a product that is fit for use. Success is measured first of all in physical terms — conformance to specifications, maintainability, etc., secondly in terms of customer satisfaction, and only thirdly in economic terms. On the other hand, no department can be allowed unlimited expenditure to achieve its quality standards. Every business expenditure must be related ultimately to a source of income. A quality plan that indulges in disproportionate expenditure, for whatever reason, is described as 'perfectionist'. This is usually caused by engineering perfectionism, but there are other ways it may come about. Polycell's division of the filler market in the 1970s into ever more refined types of product, in pursuit of their policy of market segmentation, meant that in the end they were producing a product far more specialised than the ordinary customer wanted. This left them vulnerable to being undercut by a less specialised 'all-purpose' product, which they ultimately were. The new product rapidly took a major share of the market.

It is the responsibility of senior management to control the growth of the company, and the common denominator they use is money. Ultimately the directors are responsible to the shareholders, to ensure a good return for the cash they have invested in the company. The first and last consideration at this level therefore is return on investment. To achieve a good return, however, profit margin on sales and asset turnover have to be considered, which in turn depend on number of sales, cost of sales, and assets employed; all these factors are top management concerns, and expressed in money terms. At middle management level it is necessary to talk a language which mixes money considerations with physical. Production managers are concerned both with the control of costs and with physically making the product. At the level of production supervisor the concern is almost wholly with the physical problems and quantities. A supervisor may well have an extremely accurate idea of the total production output of the company in quantity, but most probably will have no way of turning that knowledge into an idea of profitability. As we go up the company, the language becomes progressively more removed from the physical realities, and progressively more concerned with the financial implications.

Since quality is primarily a physical attribute, it is discussed and achieved in terms of physical measurements. The language used is the language of things. However, it is necessary that its operations be expressed in terms of cost for the benefit of senior management. The quality programme will affect a company's economics in two ways: it will have an effect on income, and it will have an effect on costs.

Quality and income

Recent research suggests that high product quality relative to the competition is one of the most important factors in increasing return on investment and cash flow. This conclusion is based on information gathered from over 3,000 business units across the world by an institute attached to the Harvard Business School. This database was collected to examine the *Profit Impact of Market Strategy* (PIMS). The collection of information was started in the late 1950s by General Electric to help them decide on investment and development programmes. Its purpose is to discover what makes some businesses successful and others unsuccessful. To do this, the programme identified two hundred possible factors such as return on investment, cash flow, market share, degree of unionisation, research expenditure, assets and industry growth. The impact of each of these factors on business success was then assessed. The research isolated thirty important factors in profit growth, of which four are critical. The most important is market share. It is no surprise either to businesspeople or economists that the more of a monopoly a business has, the more profitable it is likely to be. Unfortunately few Irish companies are likely to be able to achieve such a position. It is, for instance, English book publishers who have the monopoly on Yeats, Joyce, Edna O'Brien and countless other Irish writers.

The second most important factor is relative product quality (RPQ). The method of calculation of RPQ was described in Chapter 2. The critical points of the concept are that it describes:
— the customer's view, not the company's; it is not primarily therefore a matter of meeting technical specifications, but of providing fitness for use;
— both the product *and* the associated services as a package; the customer wants the brilliant engineering, plus the manual, plus ready access to the service engineer;
— quality relative to the served marketplace; a family saloon is not to be compared in absolute terms to a Mercedes — in relative terms it may even score higher!
— everything about the product *except* price.

The findings of the PIMS database show that companies with a high RPQ score a consistently high return on investment under all sorts of economic conditions. Detailed findings suggest that companies with a high RPQ achieve very nearly as high a return on investment as those with high market share. Companies with a high RPQ and a modest market share scored a higher return on investment than companies with a moderate RPQ and a high market share. High expenditure on marketing does not compensate for inferior or even moderate RPQ. Other findings of the PIMS data confirm that higher quality businesses did not have higher relative costs, but even so were able to charge premium prices. Quality control helped cost control. Companies with high relative product quality were also better able to resist outside economic pressures. Their return on investment was, in high inflation times, double that of companies with middling RPQ. In periods of severe recession high RPQ companies scored over twice the return on capital of low RPQ companies.

The conclusions of the PIMS programme are based on the results, both good and bad, of over 3,000 businesses. They confirm the experience of Irish companies such as Standex Ireland of Co. Laois, a rotary vane pump manufacturer employing 34 people, which announced a 60 per cent increase in sales after a year long Company Wide Quality Improvement Programme.

Quality and costs

The other way in which quality can affect the economics of a company is in cost. Quality costs are defined as the costs resulting from the making or supplying of defective or unsatisfactory products. In an ideal world all design and manufacturing processes would be faultless, and there would be no quality costs. In our world, the costs of ensuring the quality of products varies from as little as 2 per cent of sales to as much as 20 per cent. The average is 10 per cent of sales. This means that Irish industry spends over £1,000 million every year on quality costs.

A very low figure for quality costs generally implies that the gathering of information is faulty in some way. One management consultant specialising in quality estimates that most managements without special quality costing programmes underestimate their total quality costs by as much as two-thirds. This can happen if quality isn't perceived by management as a single concept. The cost of quality is much more than just the cost of the quality control department. Very often relevant costs are charged to other departments of the plant. Because accounting systems are not designed to gather this information, it can be quite difficult to get a true and full quality cost figure.

A complete quality cost system should include items from the following four categories:

— *prevention*: the costs of trying to ensure that we do it right the first time, typically 5 per cent of total quality costs;
— *appraisal*: the costs of checking to see whether we actually did do it right the first time, typically 35 per cent of quality costs;
— *internal failure*: the price we paid when we found out that we didn't do it right the first time, typically 40 per cent of quality costs;
— *external failure*: the price we paid for failing to discover that we didn't do it right the first time, typically 20 per cent of quality costs.

The first two categories can be described as the investment required to improve quality. The second two, internal and external failure, are the costs of inadequate quality assurance. The last category includes only the measurable costs from repair and rework. It cannot include anything for the loss of future sales to the dissatisfied customers. The various elements that make up each of these categories are listed below. The list is based on IS 304: 1984 and BS 6143. The categories are recommended by these Standards as providing a framework for a regular company quality cost report.

1. *Prevention costs*: These are the costs incurred in planning, implementing and maintaining the quality system. This includes quality work done by people not on the quality management payroll. Thus if a proportion of a supervisor's time is allocated to quality over-viewing, then that proportion of his or her salary should be allocated to the quality cost of the product. Prevention costs can be divided into several categories. Some companies allocate specific items of quality cost to different categories. For instance, certification costs and some auditing costs are often allocated to appraisal rather than prevention. Such differences between companies can restrict the comparability of cost percentages. The costs are:
— *quality engineering*: divided into costs associated with planning the quality system and translating design and customer quality requirements into controls on materials, processes and products — costs incurred in defect prevention; the second set of costs consists of those incurred in implementing the quality plan to prevent failures.
— *development of quality measurement and control equipment*;
— *quality prevention input by staff not on the quality assurance payroll*;
— *calibration and maintenance of test, inspection and production equipment*;
— *supplier evaluation and control*;
— *quality training*;
— *administration, audit and improvement* unless included elsewhere.

2. *Appraisal costs*: These are the costs incurred to discover the condition of the product, and to assure conformance with specification. They include:
— *laboratory acceptance testing*;
— *quality assurance inspection and test*: including any setup time or cost, and any materials consumed or destroyed in testing;
— *inspection by staff not on quality assurance payroll*: the costs of sorting rejected lots should go under internal failure;
— *product quality audits*: including outside endorsements and approvals and special product evaluations;
— *miscellaneous testing costs*: including costs for field performance testing and processing test reports.

3. *Internal failure*: If no defects existed in the product, there would be no failure costs. As it is, they should be listed under the following cost headings.
— *scrap*: the net loss in materials and labour resulting from manufactured goods that cannot be repaired or re-used;

— *rework*: the cost of correcting defectives to make them conform. Includes reinspection and retest costs, and the cost of appraising whether materials can be reworked;
— *down-time*: the cost of idle facilities resulting from defects such as the stoppage of a printing machine because of poor quality paper;
— *yield losses*: the cost of process yields lower than specification because of machine variability, e.g. the cost of over-filling packages. Includes loss of income from selling downgraded products as seconds.

4. *External failure*: The last category includes costs arising from the response to customer complaints. These include:
— *complaint adjustment*: all the costs of investigating and adjusting customers' complaints;
— *returned material*: costs associated with the return of faulty materials;
— *warranty charges and allowances*: costs involved in service to customers under warranty contracts and other non-contractual concessions.

Controlling quality costs

Once quality costs have been identified, they become subject to control. The objective is to obtain the desired 'fitness for use' while incurring the minimum quality costs. Another way of expressing this is to ask — what are the right quality costs?

Apart from some general rules of thumb, there are no hard and fast guidelines. This is because the requirements for quality differ too widely between, say, firms making scrubbing brushes and firms making surgical equipment. In high volume capital goods industries for instance, the quality costs are normally between 5 and 8 per cent of sales; in precision components, 7 to 10 per cent; and in high precision goods as much as 15 per cent. In consumer products industries it has been found that failure costs (internal and external) generally run at several times the appraisal costs. Another rule of thumb is that prevention costs commonly run under 10 per cent of all quality costs. Although these are vague guidelines they do open the door to an analysis of the optimum quality cost via relationships between cost categories.

The diagram on page 226 shows that prevention and appraisal costs are increased to reduce the percentage of defective items. No amount of prevention and appraisal, however, will produce 100 per cent conforming products. Because of this, there will always be failure costs. As expenditure per unit on prevention and appraisal goes up, the frequency, and therefore the cost, of failures goes down. If the two values are added together for each level of conformance, a total quality cost curve can be drawn. This has an egg-cup shape. The lowest point of the egg-cup is where total failure cost plus total prevention cost is least. The total quality cost curve can be divided into three segments. The left-hand

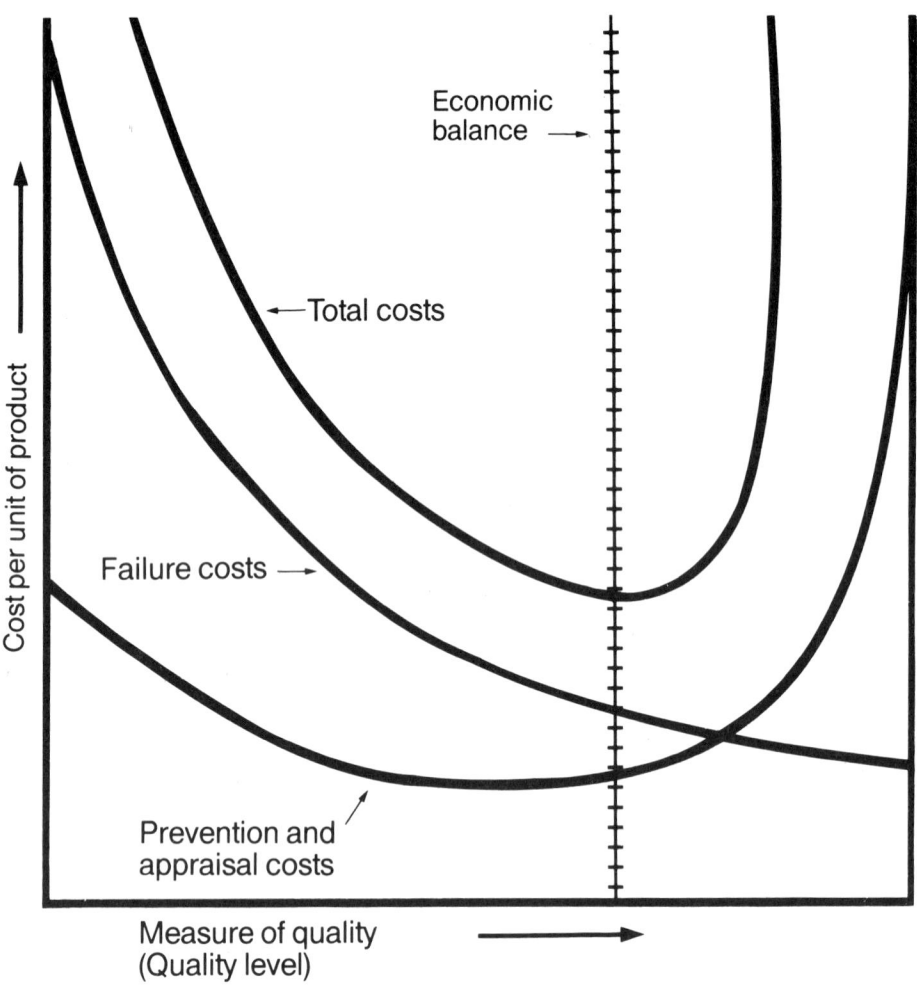

The quality costs model

segment represents the state of affairs before a quality programme has been started. Failure costs represent more than 70 per cent of quality costs, and prevention less than 10 per cent. Experience has shown that by doubling the prevention costs from this position, the failure costs can be halved. The total quality costs then drop, as shown on the quality cost curve. In financial, as well as medical, terms an ounce of prevention is worth a ton of cure. This action will tend to bring the total quality cost nearer to where failure costs are about half of total costs. If failure costs go down still further as a percentage of total quality costs, and prevention and appraisal costs continue to rise, the company is probably moving into the zone of perfectionism.

In the early stages of a quality programme it may not be possible to include all the elements in the first reports of quality costs. However, the first step towards attacking these costs is to have some indicator to measure success.

Exhibit: A quick method of producing a quality cost index
1. Select a time base, such as a month or a quarter. All the figures should be available for this period.
2. Find the cash value of saleable production.
3. Find the cash value of scrap, assuming that it had been saleable.
4. Estimate direct labour costs of inspection, quality control and reworking and add appropriate overhead charges.
5. Obtain the quality cost index by 100 + 100 (scrap value (as in 3) + inspection and reworking (as in 4))/(saleable production (as in 2)).

The next step is to address the largest element of quality costs. This will probably be failure costs. Using the tools of problem solving discussed in Chapter 20, especially 80/20 analysis, the most glaring problem should be tackled first. The cause of the specific failure should be analysed, and the corrective action specified. Preventive techniques as well as inspection and test routines should home in on this particular fault, until its reduction has become part of the routine. The quality team then tackles the next most significant problem. Gradually the structure of total quality costs shifts, as more is spent on prevention and appraisal, and less on failure.

Perhaps the most striking effect of this exercise is that not only does the output quality rise, thus securing benefits in the marketplace, but total quality costs also fall. As one writer put it: *Quality is Free!*

Chapter 26

Getting close to the customer

EVERY year senior executives at Disneyland don 'theme costumes' (e.g. Mickey Mouse suits), and for a whole week sell tickets or popcorn, dish ice-cream or hot dogs, take tickets for rides, drive the monorails, or take on any one of the jobs dealing with customers in the entertainment parks. Customers, incidentally, are always referred to in company literature as Guests.

This is worth doing, because it is easy for any organisation to forget why it's there. As one American business writer put it, 'in too many companies, the customer has become a bloody nuisance whose unpredictable behavior damages carefully made strategic plans, whose activities mess up computer operations, and who stubbornly insists that purchased products should work'. People slip into that attitude very easily. After all, they work daily with their colleagues, whose problems naturally loom larger than those of the often anonymous group of customers.

Quality is defined as fitness for use. This means that a product that leaves the factory gate having passed all the specification tests must also seem perfect when it's being used. Manufacturers claiming a quality image must continuously explore the customers' attitudes to and expectations of the product. In terms of the relative product quality concept, they have to discover their 'served market'. The various techniques for exploring this market are collectively called 'market research'. This is the study of the requirements of various markets, the acceptability of products and methods of developing new and existing markets.

Various techniques are used in market research. At the simplest level, sales staff may provide information, and past sales are projected forward. More sophisticated techniques involve the use of consumer surveys and other statistical techniques, backed up by desk research. Formal market research of this sort dates back to the 1920s in Germany and the 1930s in Sweden and France. After the Second World War, the US led in the use and refinement of market research techniques. Now market research is used even in Communist countries. In 1965 the Soviet Union set up a Market Research Institute to study long-term trends in consumer interests and expenditures. The *Administration Yearbook and Diary*, published by the Institute of Public Administration, lists eleven firms in Ireland specialising in market research.

For most companies market research is used to answer three types of question:
1. Who does (or might) use the product, when, where and why?

2. How does (or will) the product relate to the competition in terms of perceived quality and price?
3. How successful is the physical distribution of the product?

Customer surveys

When a company designs a product, and writes specifications for it, there is usually a clear use-model in view. The time, the manner and the conditions in which the product is to be used are clearly envisaged. The specifications are controlled by this model. Quality control and marketing seek to maximise the benefits the customer gains, assuming that the use-model is correct. But suppose it isn't? After all, a lot of people use toothbrushes for cleaning intricate things such as silverware. Once that question has been asked, two problems arise: how will we find out, and what can we do about what we find out? The answer is nearly always some form of customer research.

The best known research technique is the consumer survey, which can be organised for a particular purpose, or can be a regular 'omnibus survey'. In the latter case the research company organises the sample and the field research work and companies pay so much for each question they wish to include. This technique enables companies to get a quick answer to questions for which it would be too expensive to organise special research.

Consumer surveys apply statistical sampling theory to the problem of discovering people's opinions. We know from the central limit theorem (see Chapter 21) that the average of a sample, whatever the shape of the graph of the data from which the sample is taken, will in all probability be near the average of the underlying data. If we were to take a series of samples, the graph of the sample averages would be normally distributed around the population mean. The same applies to sample proportion, as in the case of trying to discover the proportion of people in Dublin who would vote Fianna Fáil if there was a general election tomorrow. As a result of this theorem, we don't have to worry about the underlying distribution when we take samples.

Two major principles underlie all sample design, whether in the market or in the factory. The first is the desire to avoid bias in the selection procedure, and the second is the need to achieve the maximum precision the budget will allow. In market research bias commonly arises from the use of non-random ('quota') samples, from errors in the sampling frame, and from refusal of people to respond.

Quota, or judgement, sampling is used in market research mainly because it is cheap and administratively easy. It can also be the only way to get a very quick response (e.g. to a television advertisement the previous night). The interviewers are given a questionnaire and a quota list. This details the number and type of people they must get to answer the questions. Thus the quota for an interviewer might consist of 20 people, broken down as 10 male, 10 female; 5 each in the age groups 20-29, 30-44, 45-64, 65+; 2 higher class, 5 middle class and 13 working class. The interviewer then has to fill the quota by picking those people passing on the street who appear to fit the categories to complete the questionnaire. Not surprisingly quota samples tend to be biased towards the better educated, the more extrovert (or approachable) types, and towards certain occupations.

A random sample design is one which gives every unit in the population a definable chance of being selected. In order to do this, however, the researcher must have a sampling frame. This is the list, index, map, etc. from which the sample is to be drawn. Without a sampling frame, it is impossible to plan a sampling technique that is random. Unfortunately sampling frames have their own problems. For instance, telephone directories have often been treated as a sampling frame that would at least take in the richer half of the population: however a stockbroker recently reported that a quarter of his private clients were ex-directory. Other problems that can arise include blank elements (e.g. where the subscriber has moved), clusters (e.g. where one number serves a large hostel) and duplicate listings (e.g. households with more than one telephone line). The man who was told that every fourth child born in the world was Chinese, so decided to stop after his third, had sampling frame problems.

Once the sample is drawn, the next step is to construct the questions to be asked. This is a delicate business. Books have been written covering such matters as the level of language to be used, ways of avoiding questions that suggest an answer, the best order of questions, and the maximum length an interview should be.

Exhibit: Market research questions not to ask
1. Vague questions: 'Are you satisfied with your canteen?'
What do I answer if I like the food but hate the noise or the decor?
2. Complex questions: 'Do you think the government is justified in pursuing its current policies of monetarism, given the present trend away from supply-side economics?'
What?
3. Ambiguous questions: 'Do you read books regularly?'
For many people books include magazines; regularly might mean every Christmas Day.
4. Leading or presuming questions: 'Do you think you should give up cigarette smoking?'
Or the lawyer's favourite leading question: 'When did you stop beating your wife?' The only way to answer this is to step outside the framework of the question altogether, which is sometimes difficult to do.
5. Personal questions, bluntly phrased: 'Did you buy a toothbrush in the last three months?'
A question similar to this in a British consumer survey in the early 1970s was answered so overwhelmingly positively as to imply toothbrush sales some five times larger than they could have been — unless there was a big second-hand market.

Finally the questionnaire designer should be careful of the order. Respondents put a higher priority on presenting a coherent image to the interviewer than a totally truthful one. The ordinary muddle of daily life tends to be suppressed in favour of apparent logicality. In the pursuit of this logical self-image, the answer to one question can bias the answer to a subsequent one. The lady who was asked first: did she have children? and then: was she married? had reasonable complaint.

As the interviewer runs down the questions, the form is ticked, very often on pre-coded boxes. This enables the data to be analysed very quickly, though open (uncoded) answers which are coded in the research office can sometimes be more revealing. The data is then analysed according to ordinary statistical techniques. Part of the market researcher's importance to the client, however, lies in the interpretation put on the data. The impact of difficulties in the sample needs to be interpreted with an eye to the realities of the marketplace, as do the statistical tolerances involved.

Panel techniques

Survey techniques have two problems: firstly, people tend to simplify and exaggerate their behaviour patterns. They do this to appear credible, but also simply not-illogical. Secondly, surveys provide good snapshots, but are less satisfactory as indicators of trends, because of the difference between the samples. To get over these problems, market researchers invented the panel technique, which is based on the idea that actions speak louder than words.

A consumer panel consists of a statistically selected group of households which record their weekly purchases of a group of products. In some cases they fill in a preprinted diary, in others a research field worker counts the boxes in the store-cupboard and the empty wrappers and cartons in the special bin provided. Sometimes an interview may be added to the weekly audit visit. One British panel has been going for more than twenty years and now has 5,000 households reporting to it. This size enables manufacturers to monitor purchasing behaviour very closely. Such information as brand loyalty, response to promotions, take-up of new products and seasonal purchasing patterns for a whole range of products can be precisely monitored. An important result of these panels is brand share information. Short-term panels can be set up to reflect the success of a test marketing exercise, or a special promotion. A special development of the consumer panel of interest to quality students is the *disposition survey*. This examines what has happened to durable purchases such as TV sets. There are three options: it can be kept, rented or loaned elsewhere, or disposed of. If kept, is it being used as was intended, or simply stored, or used for some other purpose? And so on. Clearly it is of great interest to quality engineers to discover how their products end up, and how long they remain in full use.

The special problems associated with consumer panels are related to the technique. Less than half the households approached actually agree to come on to the panel (despite inducements); the drop-out rate is about 25 per cent. Particular goods, such as tobacco, shampoo, DIY equipment or consumer durables are not well recorded because they are bought as individual or infrequent purchases, rather than as part of the normal household purchasing stream. Particular types of household, such as the very rich, the very poor and single person households are badly represented on panels.

Despite these limitations, consumer or household panels can enable the manufacturer to monitor how sales are doing against rival products, and also, through interviews, to gain an insight into the final use of the product. But an important part of the quality chain is the distribution system, which carries the product from the factory to the consumer. Because the manufacturer has little knowledge or control over what happens to the product once it leaves the warehouse, special market research techniques are used to monitor this area. The main one is a panel technique called *retail auditing*. It is based on the simple theory that the sales of any one product through any one shop can be discovered by the following sum:

Stock at beginning of period *plus* purchases during period *less* stock at end of period *equals* sales during the period.

Since the stock at the end of period one is the same as the stock at the beginning of period two, only two figures need to be discovered every period. Occasionally it may be necessary to adjust the figures for breakages or theft. Assuming the retailer permits it, there is nothing to stop a researcher from counting the stocks and purchases of several products from different manufacturers. If the stocks and the purchases are counted for a range of products, say cigarettes, it becomes possible to calculate the relative sales of various brands through the shop. Extend the idea to a large sample of shops, and we have a technique to enable a company making 'irregular-purchase' items such as paint, or non-household items such as tobacco, to gain the information that the consumer panel couldn't give.

Not only can the company find out the brand share of its product and its rivals', but it can also learn about how the retailers' purchases and stocks are moving. This can be important tactical information. There is no point in launching a big consumer promotion, for instance, if retailers' stocks are very low. Stocking and distribution information also enables the company to monitor how long, on average, stocks are kept on retailers' shelves, and how this changes over the year. It is important that the product be in perfect condition when the customer uses it. If retailers are keeping stocks too long, perhaps because of bulk buying incentives, delivered quality will fall.

Exhibit: Launching Yoplait

Yoplait yoghurt was launched in 1974, and today holds in excess of 75 per cent of the market. The total yoghurt market has grown in that time by 300 per cent. The company has used market research continuously during that time to monitor its progress.

The first research was undertaken in 1972, when it was discovered that 86 per cent of households never bought yoghurt; only 9 per cent bought regularly. The market was biased towards ABC 1 Dublin households with children. After a series of central location tests comparing recipes, the manufacturers found to their surprise a bias away from French gastronomic expertise towards an Irish image, stressing the quality of Irish dairy products.

The launch was specifically aimed at ABC 1 housewives under 45 with children, with a particular emphasis on Dublin and Cork. Single females were an important second market. The two objectives of attracting new users and urging existing users to switch were successful, and resulted in a 25 per cent share of the market within four weeks.

Post-launch research was based on omnibus surveys, which reported a satisfying growth in two years from a mere 14 per cent of regular housewife users to 60 per cent.

By 1977 the company began to worry about the market plateauing, so it commissioned some qualitative research to examine the general market situation. This highlighted the increasingly cosmopolitan attitudes of Irish housewives to food, and the general use of yoghurts as desserts and snacks. The nutritional goodness of yoghurt was strongly in the consumers' minds.

Eight years after the original launch, the product was relaunched, following research into purchasing patterns. This discovered, for instance, that 93 per cent of purchasers bought more than one pack at a time. As a result of this discovery, a new multiple pack was launched, which eventually increased sales volume by 10-15 per cent.

The company's experience has been that if one intends to maintain the quality the customer wants, one must continually evaluate the product, the market, and in particular the performance of the product in the marketplace.

Source: Extracted from a paper by Brian J. Milton, November 1981

The technique of retail auditing is usually based on a panel of shops selected on a quota system. The researcher estimates the numbers of shops in each category in different areas, and models the panel accordingly. The reports are broken down by brand, by area and by type of outlet (e.g. pub, newsagent, large or small supermarket, etc.). This information enables the company to check that distribution is efficient to all areas and types of outlet.

In-depth research techniques

A large number of more or less exotic techniques have been devised, loosely based on psychological theory, to explore in detail consumers' feelings and thoughts about products. These are based on non-randomly selected groups or individuals, who might engage either in a loosely directed discussion or a brainstorming session about the product. These can of course be recorded on video. Another approach is to expose a few people to in-depth interviews. Projective techniques involve the researcher presenting some stimulus to the subject, for instance word association tests, sentence completion (I like Baileys because ...), fantasy exercises (imagine you are a lawnmower — what do you feel?), cartoon completion, or interpretation of an ambiguous picture involving the product. Role rehearsal involves the subject being asked (and paid) to do something slightly out of the ordinary in order to stimulate a flow of response to the product: one housewife was asked to serve her family chicken three days a week for a year, and to record their remarks!

Other consumer research techniques

The number of consumer research techniques is limited only by ingenuity and budgets. In the early 1960s, for instance, much was made of a technique for filming the eyes of supermarket shoppers, unknown to them of course. It was believed that the involuntary flickering and blinking would reveal much about response to packaging, shelf filling and other matters. More seriously, many techniques have been devised to report on the attributes of products.

Most familiar of these is the *central location product test* (also called town hall product tests) in which people are drawn off the street according to a quota and asked to taste a group of products and make some judgement. A more sophisticated version of this is the quality panel. This technique involves giving a product to a sample of consumers for judgement. The sample is selected by quota according to whether they are heavy or light users, and by demographic factors. The purpose is to enable the manufacturer to monitor the consistency of the quality of the product. Unsatisfactory samples are returned to the client for further analysis. Another use of the panels is to monitor response to potential variants of the product.

Industrial market research

Industrial market research, or research into behaviour patterns among industrial buyers, is not as well developed as consumer research. There are special problems in confronting professional buyers, particularly since there will usually be at least three influences involved in the purchase. These are the purchasing agent, the engineer or design consultant, and the user. Each of these has a different set of objectives to fulfil. A further problem is that industrial marketing is very fragmented. Most sellers have a relatively small list of potential purchasers (about whom they are inclined to be secretive). To some extent this problem can be met through trade associations, but many of these are cockpits of extreme mutual suspicion. This makes statistically oriented research techniques very difficult to apply.

Conclusion

The book *In Search of Excellence* by Peters and Waterman stressed the necessity for companies to keep close to the customer. There is a great temptation to become wound up in what is happening in-house, and ignore the real world out there. Market research is one of the best ways in which this can be done, especially when it is related to other sources of information about the company's customers, such as the complaints procedures and the reports from retailers and salespeople.

Market research is paid for by people who want to increase sales. It is generally therefore slanted to the manufacturer's interest. If, for instance, you ask a fridge company why fridges are always white, they will answer that sales prove that customers want white fridges. Furthermore, the fact that only white fridges are available proves there is no other demand, because if there was, coloured fridges would be available. Perhaps there would be no demand for coloured fridges. But when only white bathroom furniture was available, no doubt manufacturers said the same. Economists distinguish between vague desires ('I'd like a villa in Portugal') and effective demand (i.e. something that can be paid for now). The traditional theory was that any effective demand would automatically, in the jostling of the marketplace, create supply. In the 1960s, consumers began to notice that large corporations had a financial interest in limiting the diversity of certain markets. A proportion of the money spent on advertising, for instance, is spent not primarily to promote the goods concerned, but to make it too expensive for potential competitors to enter the market. This kind of realisation led to the birth in the 1960s of the consumer movement.

Chapter 27

Consumer protection

The consumer movement

In the 1960s there grew up, in several industrialised countries, a vivid realisation that a few large corporations controlled much of the market in consumer goods. Nowadays, for instance, some six retail groups sell over 75 per cent of groceries sold in Britain. In Ireland the markets for electricity (the ESB), public transport (CIE), cement (Roadstone), beer (Guinness), spirits (Irish Distillers) and sugar (Irish Sugar) are only a few with dominating companies. In 1971 multiple stores sold 50 per cent of groceries in Dublin and 30 per cent in the country. The figures in 1985 were 81 per cent and 58 per cent.

The PIMS database confirms that nothing, not even product quality, gives a better corporate return on investment than a large market share. Companies therefore spend much of their effort in trying to gain a dominating share of the market. Furthermore, the larger the market share, the more effectively companies are able to flout the traditional assumption of market theory, which is that the consumer is king. Companies in this position are able, for instance, to have a 'take-it-or-leave-it' attitude to product quality. The consumer movement, and the legislative response to it, attempts to tip the balance back in the customers' favour.

When General Motors launched the rear-engined Corvair in the 1960s, they were told that the rear wheels had a fatal tendency to 'tuck under' on bends. One of the early casualties of this car, which a young law student called Ralph Nader described in the title of his famous book as *Unsafe at Any Speed*, was the son of one of the top GM executives. Nader's book was reprinted; he was called as a witness to a US Senate committee on motor car safety. Nader was harassed by General Motors, a fact which he rapidly made public. This attempted bullying did more to impress the American public of the potential irresponsibility of large corporations than any amount of defective goods could have done. The fact that the offender was the company about whom it had been said 'What's good for General Motors is good for America' only rammed the point home.

Nader went on to produce a series of reports about American cars, and to set up various consumer organisations. One of his reports described the bumpers of a certain make as 'designed to withstand collisions of up to 2.8 mph'. He went on to point out that 'this is calculated engineering design to increase sales of spare parts, giving General Motors more bucks for the bang'. This aggressive approach did not endear 'Nader's Raiders', as they called themselves, to the industrial leaders of the nation. They were described as the 'most dangerous men and women in America today'. But now, partly as a result of Nader's activity, the US has the most stringent product liability law in the world.

In Britain, the main impetus came from the Consumers' Association, which had been set up in 1957 to publish comparative product use reports. The first issue of their magazine *Which?* covered kettles, cake mixes and cleaning powders, and published a Swedish report comparing the Austin A35 and the Standard 10. It wasn't much more popular with the business establishment than Nader: 'it undermines consumer confidence', said one spokesman, 'it is a very bad factor in the present economic system'. The Consumers' Association now has over 625,000 subscribers to its various publications. The Irish version, the Consumers' Association of Ireland, was set up in 1966.

Exhibit: The attitude that led to the consumer movement
British Rail had just completed a splendid new terminus for one of its most prestigious routes. The architecture, the timetable displays, the cafés, the ticket hall, the bookstalls were all in the best possible taste. The station won awards. The only cloud on the horizon was spotted by the woman who asked why there were no benches for the passengers to sit on. 'Well,' explained the BR spokesperson, 'we decided not to have benches because people make such a mess with litter while they sit on them.'

The Consumer Information Act was passed in Ireland in 1978. It set up the Office of the Director of Consumer Affairs, whose function was to monitor the accuracy and reliability of information supplied to consumers in advertising, promotion and pricing of goods and services. The Office's 1985 annual report notes that they received some 13,000 complaints in 1984. By far the largest categories were complaints about clothes and cars; electrical equipment and furniture and household fittings came next. But these complaints are only the tip of the iceberg. Research carried out for the Office in 1982 discovered that only 3 out of 1,114 people with a complaint actually consulted a lawyer about it, and only 2 contacted the Office of Consumer Affairs. Three-quarters of those who complained to the company concerned were unsatisfied with the response.

From the point of view of the quality function, the consumer movement represents feedback from the customers. Quality after all is commonly defined as 'fitness for purpose'; the people who discover in detail how a product really works are those who use it day after day. Further, they can make clear to the manufacturer which aspects of the product they find important and which merely decorative.

Consumer protection in Ireland

To protect their rights, consumers can resort to common law (i.e. traditional case law), to statute law and to the organisations set up under statute, and to consumer organisations.

Common law consumer protection: The common law takes two approaches to the question of protecting the consumer. The first is through contract. What does or does not constitute a contract can be a highly technical question, but in principle a person can enforce a promise if:
— there is an agreement intended by both parties to be legally binding;
— there is some form of consideration (e.g. payment);
— the parties are legally entitled to contract (e.g. they are not under age);
— the contract is not illegal, unenforceable, void or voidable.

If these conditions are met, and in most cases they are, then the contract will be enforceable in the courts. The most common form of contract is one for sale of goods. This has been covered by legislation since 1893.

The second way the common law protects the consumer is through tort. Tort is a Norman French word meaning 'wrong'. The law of tort covers wrongs done by one person to another outside contractual relations. In general, only specific types of 'wrong' are covered by the law. However, by the English leading case of Donoghue v. Stevenson (1932) (which was received into Irish law through Burke and Holloway v. Kirby (1944) IR 207) the manufacturer also has a 'duty of care to the ultimate user of the product'. In Donoghue v. Stevenson, two old ladies had stopped at a café for refreshments. The one standing treat bought an ice-cream for her friend, and then some ginger beer. The friend drank some of the ginger beer, and then the rest was poured over the ice-cream. As the liquid was poured out of the opaque bottle, the decayed remains of a snail slithered out of the bottle on to the ice-cream. Naturally the old lady was very upset. She later alleged that she had contracted a serious illness as a result. However, since she had no contractual relationship with the café-owner, nor of course with the manufacturer, she had at first no remedy. However the House of Lords decided that the manufacturer or supplier is bound to take reasonable care to see that their goods and services are reasonably fit for their purposes. The catch is that the injured person has to prove in court that the supplier did not take reasonable care, which can be very difficult. Recent EEC product liability legislation, which will be Irish law by 1988, and which is discussed in the next chapter, changes this burden of proof completely.

Statutory consumer protection: The *Sale of Goods Act, 1893*, as amended by the *Sale of Goods and Supply of Services Act, 1980* is the primary piece of legislation attempting to protect the consumer, moving away from the old legal maxim *caveat emptor*, which means 'let the buyer beware'. This Act laid down that the purchaser can expect goods to be
— of merchantable quality;
— fit for their normal purpose unless the defects were pointed out at the time of purchase;
— reasonably fit for the purpose for which the buyer intends them: a bun with a needle in it may be fine as a pin-cushion, but not as a snack;
— they must also be as described on the package or in advertisements, or if bought from a sample must conform to that sample.

The responsibility for obeying the law rests almost always with the seller. If anything goes wrong, the retailer has a duty to set it right. What is more, it is an offence for the retailer to imply that the consumer does not have these rights. As a result statements such as 'No liability accepted for faulty goods' or 'No money refunded' are actually illegal. If the goods are faulty, the consumer may be entitled to reject them, to get a partial refund, or to agree provisionally to a repair. If the repair isn't satisfactory, the consumer may reject the goods. Services are also covered under the Act. The consumer has a similar right to expect that the supplier has the necessary skill, that the materials used are of merchantable quality and that the service will be provided with due care and attention.

The *Industrial Research and Standards Act, 1961*, as well as reconstituting the Institute for Industrial Research and Standards, included certain consumer protection measures. It is an offence under this Act to make a false claim to compliance with any Irish Standard. A justified but unlicensed claim is not prohibited. It is, however, an offence to use a Standard Mark without a licence. Under this Act also the Minister may prohibit the manufacture or sale of any substandard goods on grounds of health or safety. A specific order must be drawn up for each type of goods.

The *Consumer Information Act, 1978* prohibits false or misleading advertising and manufacturers' and traders' claims. The two main offences it creates are:
— to apply a false trade description to goods;
— to sell or possess for sale goods to which a false trade description has been applied.

The Act details twelve different categories of false description, ranging from 'number, quantity, measure, gauge, capacity or weight of any goods' to 'any physical characteristic not previously referred to' and 'as to the conformity of any goods with any standard or their passing of any test or their commendation by any person'. Thus such descriptions as: 'guaranteed Irish', 'rustproof', 'home-made', 'conforms with IS 300', 'length — 20 metres', 'written by' can all be offences against the Act. A major category of misleading or false statement is that referring to price. Misleading indications of present, previous or recommended prices are all prohibited. Statements about services are also covered in the Act, but the burden is slightly different. In this case it is an offence to give, knowingly or recklessly, false or misleading indications as to services provided in the course of a business, trade or profession. Statements such as: 'Delivery to all areas', 'No job too small', '24-hour service', 'All makes serviced' are all covered by the Act. For statements about services, there is no offence if the statement was made in good faith and the person making it had good reasons for believing it to be true.

The Director of Consumer Affairs is charged with enforcing the Act, which carries penalties of fines up to £500 or imprisonment for up to six months or both. In some cases goods may even be forfeited. Once an offence has been officially notified, the Director will typically 'request' the offender to refrain from deceptive trading practices. If such a request is not complied with, the Director may seek a High Court order. The Director's Report for 1984 records 541 complaints of false or misleading descriptions. Prosecutions were considered in twelve cases, six of which related to motor cars and four to holidays misdescribed in brochures.

Consumer protection organisations

A wide variety of organisations provide some information or protection to the consumer. Examples are:

— *The Director of Consumer Affairs*: soon to be Director of Consumer Affairs and Fair Trading
— *Department of Health*: Health Inspectorate
— *Department of Industry*: Safety Inspectorate
— *Advertising Standards Authority*: a voluntary self-regulating body set up by the advertising industry to monitor contraventions of the Code of Advertising Standards.
— *Consumers' Association of Ireland*: founded in 1966 as a wholly independent association to protect and educate consumers in Ireland. Publishes *Consumer Choice*.
— *Irish Housewives' Association*: founded in 1942 as a voluntary organisation to protect and defend consumers' rights as they are affected by the supply, distribution and price of essential commodities.

The EEC and consumer protection

In 1975 the EEC proclaimed that it would support five basic consumer rights. It works towards these goals by producing general rules for all members. These *Directives* have to be made part of the legislation of each country within a certain time. Not every country, however, is as conscientious as Ireland in fulfilling this requirement. There are two types: Directives related to specific groups of products and general Directives. Typical Directives include:

— Directive 77/728/EEC of 7 November 1977 relating to the classification, packaging and labelling of paints, varnishes, printing inks, adhesives and similar products
— Directive 79/693/EEC of 15 January 1980 relating to fruit jams, jellies, marmalades and chestnut purées
— Directive 79/530/EEC of 14 May 1979 on the indication by labelling of the energy consumption of household appliances.

Examples of the more general type of EEC activity include the setting up in 1973 of the Consumers' Consultative Council, and the Council Decision 84/133/EEC of 15 January 1984 introducing a Community system for rapid exchange of information on dangers arising from the use of consumer products.

The five basic consumer rights are:

1. *Consumer health and safety*: The concern for consumers' safety specifically covers such products as foodstuffs, cosmetics, detergents, consumer durables, toys, etc. Various veterinary Directives have been adopted to improve health guarantees for consumers.

2. *Consumers' economic interests*: Because consumers' economic power has been eroded by large monopolistic corporations, the Community has intervened to correct the balance, particularly with proposals on such matters as terms of contract, advertising truth, labelling, after-service, tourism, and credit terms.

3. *Access to courts*: To give consumers legal rights is one thing — to make it practical for those rights to be exercised is another. The cost of legal procedures, the technical language and the slowness of many courts all serve to deter potential litigants with genuine claims.

4. *Consumer information and education*: The main improvement in this category has been in the labelling of goods. In 1978 a Directive relating to classification, labelling and packaging of dangerous substances was adopted, and in 1979 one relating to the pricing, and in 1980 one relating to the labelling and packaging, of foodstuffs. Pilot schemes and proposals are in hand for the development of consumer education in primary and secondary schools.

5. *Consumer representation*: The Community's main task has been to encourage the development of the European consumer movement through the Consumers' Consultative Committee. The CCC provides general help and advice to the Commission of the EEC, and coordinates the flow of opinion to the EEC from various national and international consumer bodies.

One recent EEC Directive will have a major impact on the quality function in every consumer industry. This is the Directive on Product Liability, which removes the need for the injured party to prove negligence in cases like Donoghue v. Stevenson. Once this is incorporated into Irish law, all the old lady would have to prove is that she was injured and that she was injured as a direct result of perfectly reasonable use of a manufacturer's product. It will then be up to the company that made the product to prove that it had not been negligent. The implications of this shift in the burden of proof are so important that they deserve a chapter to themselves.

Chapter 28

Product liability

WHILE Ralph Nader and General Motors were messily slugging it out in the public eye, over 60,000 lawsuits were filed in the US for injuries arising from the use of manufactured products. By the 1970s the total had risen to 100,000 a year, most of which were settled out of court. If that seems a lot, compare it to the US government's 1970 estimate that twenty million Americans were injured every year in incidents involving consumer goods. A recent EEC survey discovered that in the two and a half years between January 1982 and June 1984, accidents in which products were involved were the cause of some 30,000 deaths and 40 million injuries. Accidents at work and road traffic casualties were excluded from these totals. Product-related accidents accounted for more injuries than either. If the same proportion holds good in Ireland, over 350,000 people are being injured every year by consumer goods.

The US legal system responded to this by shifting the burden of proof from the injured person to the manufacturer. Once the plaintiff was able to prove that there was damage and it was caused by the product, the case was made. The available defences vary from state to state. The US Consumer Products Safety Act, 1972 set up the Consumer Products Safety Commission, which monitors the safety of products. If a company has any reason to believe that there might be a hazard to consumers from any product on the market, they must inform the Commission within forty-eight hours. The Commission will then decide what is to be done, and may order the product to be withdrawn from the market.

The product liability does not merely cover the manufacturer. Anyone in the chain of distribution may be sued. The liability depends on a defect in the product, but this includes absence or presence of directions for use, and misdirection in advertising.

Product liability in Ireland

The 1985 EEC Directive, which lays down a standard of strict product liability, must be incorporated into the legislation of member countries by 1988. The purpose of the Directive is to ensure that all EEC countries have roughly the same law on this matter, and that the law incorporates the principles of strict, or 'no-fault', liability. It is not yet known whether Irish legislature will simply incorporate the wording of the Directive into the law, or whether it will take the opportunity to make a fundamental recodification of the relevant law. As a result the exact wording of the Irish law is not yet known. Most experts believe the first course to be the more probable. The full text of the Directive is given in Appendix 1. The key sections are described below.

1. *Manufacturer's responsibility*: In order that the burden of risk shall be borne by those who can either bear it or insure adequately against it, manufacturers are to be responsible for any damage or injuries caused by their products, whether they are negligent or not. This rule applies to all industrially produced movables. This liability cannot be limited or excluded by agreement between the user and the producer.

2. *Liability*: Any company that had a part in the production shall be liable. Suppliers of components and raw materials will be initially liable for the full amount of the injury, not merely the proportion represented by their input. If goods are imported into the EEC, the importer is liable; if the producer cannot be identified, the supplier or retailer is. If a company attaches its own trade mark or identification to a product, it is deemed to be the manufacturer. Thus defects arising from own-label grocery brands will be chargeable to the retail chain involved.

3. *Reasonable use*: A product is deemed defective when it does not provide the safety which a person is entitled to expect. The circumstances to be taken into account include the presentation of the product, and 'the use which it could reasonably be expected might be made of the product'. This does not appear to be the same as reasonable use of a product. A manufacturer might have a reasonable expectation that at least some purchasers will use the product unreasonably. Most authorities, however, guided by the preamble, assume that this provision means that manufacturers have only to consider reasonable, not predictable but unreasonable, use.

4. *Defences*: Producers have some defences. Firstly, they can prove that they didn't put the product into circulation, or that it was not manufactured by them in the course of business. Secondly, they can prove that the defect that caused the damage was not in place when they sold the product. A component manufacturer can prove that the defect was caused by the product's design and not by the component, or that the component was wrongly fitted by the manufacturer. Finally, they can prove that the defect could not have been discovered at the time, given the prevailing state of scientific and technical knowledge. This of course requires that the manufacturer be familiar with the very latest in inspection and testing devices relevant to the product. The fact that the defect was caused by compliance to mandatory regulations issued by public authorities is also a defence. If the injured party either caused or contributed to the accident, that can be a defence too.

Quality in Practice

5. *Time limits*: The injured party must sue within three years of receiving the injury, and the actual product that caused the damage must have left the warehouse less than ten years before, unless an action has already been started. This means that companies will have to store batch records and quality documents for at least ten years from the date of manufacture. For products that sit in the warehouse for any time before despatch, this period should be extended accordingly.

6. *Limited liability*: The Directive states that the producer's total liability for personal injury caused by identical items with the same defect shall be limited to £50 million. To avoid many petty claims, the damage to property must exceed £350.

Implications for industry

The new law of strict liability imposes a burden on the producers of all movable products (including electricity) in respect of:
— the product itself
— labels
— packaging and containers
— installation and use instructions
— warranty documents
— sales brochures, catalogues, point of sale and other advertising materials.
Thus the product itself may be fine, but the instructions misleading. Liability can also arise if the advertising or the packaging implies some use that turns out to be dangerous.

The EEC Directive on Product Liability imposes a considerable new risk on manufacturing companies. The best way to deal with this risk is to have an effective system of quality control which will totally prevent injury or damage, thus avoiding any claims. Product safety now has an added importance in the corporate plan. Obviously the role that the quality assurance department has to play in this can be vital. Unfortunately no quality plan can be infallible, so every company will need product liability insurance.

Insurance

The first line of defence is clearly insurance. Product liability insurance, at least for firms with properly documented quality functions, will be available. This should at least ensure that damages can be paid. It will, however, be expensive. Already some companies do not have proper public liability insurance, and it is feared that many will fail to insure against product liability, thereby taking the risk on themselves of a big claim and the costs of a product recall. Insurance is already expensive in Ireland: our total premiums, as a percentage of GNP, are second in the world only to America. Germany and the UK come next. Because of the increase in claims potential, the claims severity and the legal costs that are bound to arise, insurance risks will be high. As a result insurance companies will look

carefully at the claims experience of the company and its current operating procedures before accepting the risk. Quality assurance aspects will be closely scrutinised. In particular, the insurance company will look at:
— research and design procedures
— ongoing test and inspection routines
— product labelling and instructions to ensure safe use
— exports' compliance with local regulations
— legal responsibilities

Product liability prevention

No system in the world is absolutely safe. The problem is to reduce the levels of safety to an 'acceptable' level. To do this companies must install a product liability prevention programme.

1. *Corporate safety policy*: There should be a public statement of the company's general commitment to reliable and safe products. Like the statement of quality policy in the quality manual, this should be addressed to employees and customers by the managing director. It is important that senior managers are also genuinely committed to the policy.

2. *Design and product development*: The law of strict liability puts a new focus on the design function. Products not only have to be designed with an eye to their use, but also with an eye to how customers might misuse them. Products also have to be designed and tested with the best equipment, etc. that the state of scientific and technical knowledge can supply. Further, since it is a good defence to prove that the product was thus designed to comply with mandatory national safety standards, these have to be considered. Obviously compliance with voluntary standards would be desirable as well. Because the design process effectively dictates the course of manufacturing and materials purchasing, it is the key department. Various techniques have been evolved for examining the potential hazard presented by each component, such as failure mode and effect analysis (FMEA) and fault tree analysis.

3. *Manufacturing and inspection*: Provable inspection and control of bought-in materials and the production process form an important part of product liability limitation policy. This means that the normal quality control checks and records have to be designed with a view to possible production in court. The chain of checks, from incoming materials to warehouse controls, has to be unbroken. Records of calibration of monitoring devices should be clear.

4. *Packaging, warnings, labels and instructions*: All the materials that go with the product must be designed with an idea to consumer safety. Packaging materials

Quality in Practice

must be able to withstand the temperatures (average and range) that are likely to be encountered. Conditions of shelf storage in retail outlets need to be taken into account. The labels and instructions should contain the necessary warnings. These are no substitute for safe product design, but they can at least clearly indicate the expected 'reasonable' use of the product. The user population and environment must be considered in terms of installation, use and repair information. It would be positively dangerous, for instance, to provide only a technical use and repair leaflet for an ordinary domestic product.

5. *Advertising materials*: These must be technically accurate and also accurate as to the indicated use. Point of sale materials often picture the product being used. This should not imply a type of use that might result in danger.

6. *Field monitoring*: A possible defence to a product liability suit is the so-called 'state of the art' defence. This requires the manufacturer to prove that the product could not have been known to be defective in the light of current scientific and technical knowledge. To avail of this defence, the company would have to be able to demonstrate that every care was taken with the product from manufacture until it reached the customer's hands. One useful feedback channel is customer complaints. These indicate the things about the product that the customer finds unsatisfactory, so that these matters can be redesigned before they cause damage, and also the ways in which the product is being used. If the actual use-model is in fact different from that envisaged by the design and marketing people, significant changes may have to be made to the presentation of the product. In the absence of actual complaints, a system should be devised of feedback from field service operatives and salespeople.

7. *Product recall procedures*: Even after a company has been alerted to a defect, it is under no obligation to recall the product. However, failure to do so would expose the company to severe risk of multiple cases being brought against it. An essential part of the product liability minimisation procedure is the product recall plan. This involves putting the whole distribution programme into reverse. Advertising has to announce that the product is not to be used or sold, salespeople have to remove products from shops, transport goes back laden to the factory, the accounts department sends out credit notes instead of invoices. Every single item in the marketplace has to be traced and reclaimed. This means that it has to be possible to trace the items via the quality assurance documentation of part and batch numbers.

8. *Documentation*: It is one thing to have an effective product safety policy, it is another to be able to prove it in court up to ten years hence. The US Consumer Product Safety Commission suggests that the following types of records should be maintained:

— results of all inspections, tests, calibrations
— details of consumer complaints
— design documentation, including details of any actions taken to correct product and system deficiencies
— batch or process documentation.

In fact a properly designed quality assurance system will involve keeping all these records. The major change will be in the length of time that they have to be kept, and the details of the records.

Costs

Strict product liability will cost companies money. The major category of extra cost will be insurance, but there will be others as well. The storing of quality assurance records for more than ten years implies the creation of a virtual archive. Companies will tend to move as quickly as possible to use of electronic storage and computer processing simply to get over the problem. The requirement to keep up with the very latest in scientific and technical work into possible defects will involve extra expenditure on research materials and information. More attention will be given to product design and testing, and more time and money spent on recording the various hazard test programmes. And finally, if it all goes wrong, there are the costs of a product recall. These can be enormous. When John West's had to recall one of their tinned fish products, it cost them over £25 million sterling. Their sales were still down by 25 per cent a year later.

Appendixes

APPENDIX 1

Council Directive on the approximation of the laws, regulations and administrative provisions of the Member States concerning liability for defective products

The Council of the European Communities,
Having regard to the Treaty establishing the European Economic Community, and in particular Article 100 thereof,
Having regard to the proposal from the Commission,[1]
Having regard to the Opinion of the European Parliament,[2]
Having regard to the Opinion of the Economic and Social Committee,[3]

Whereas approximation of the laws of the Member States concerning the liability of the producer for damage caused by the defectiveness of his products is necessary because the existing divergences may distort competition and affect the free movement of goods within the common market and entail a differing degree of protection of the consumer against damage caused by a defective product to his health or property;

Whereas liability without fault on the part of the producer is the sole means of adequately solving the problem, peculiar to our age of increasing technicality, of a fair apportionment of the risks inherent in modern technological production;

Whereas liability without fault should apply only to movables which have been industrially produced; whereas, as a result, it is appropriate to exclude liability for agricultural products and game, except where they have undergone a processing of an industrial nature which could cause a defect in these products; whereas the liability provided for in this Directive should also apply to movables which are used in the construction of immovables or are installed in immovables;

Whereas protection of the consumer requires that all producers involved in the production process should be made liable, insofar as their finished product, component part or any raw material supplied by them was defective; whereas, for the same reason, liability should extend to importers of products into the Community and to persons who present themselves as producers by affixing their name, trademark or other distinguishing feature or who supply a product the producer of which cannot be identified;

Whereas, in situations where several persons are liable for the same damage, the protection of the consumer requires that the injured person should be able to claim full compensation for the damage from any one of them;

---

[1] OJ No C 241, 14.10.1976, p. 9 and OJ C 271, 26.10.1979, p. 3.
[2] OJ No C 127, 21.5.1979, p. 61.
[3] OJ No C 114, 7.5.1979, p. 15.

Appendix 1: EEC Directive on strict product liability

Whereas, to protect the physical well-being and property of the consumer, the defectiveness of the product should be determined by reference not to its fitness for use but to the lack of safety which the public at large is entitled to expect; whereas the safety is assessed by excluding any misuse of the product not reasonable under the circumstances;

Whereas a fair apportionment of risk between the injured person and the producer implies that the producer should be able to free himself from liability if he furnishes proof as to the existence of certain exonerating circumstances;

Whereas the protection of the consumer requires that the liability of the producer remain unaffected by acts or omissions of other persons having contributed to cause the damage; whereas, however, the contributory negligence of the injured person may be taken into account to reduce or disallow such liability;

Whereas the protection of the consumer requires compensation for death and personal injury as well as compensation for damage to property; whereas the latter should nevertheless be limited to goods for private use or consumption and be subject to a deduction of a lower threshold of a fixed amount in order to avoid litigation in an excessive number of cases; whereas this Directive should not prejudice compensation for pain and suffering and other nonmaterial damages payable, where appropriate, under the law applicable to the case;

Whereas a uniform period of limitation for the bringing of action for compensation is in the interests both of the injured person and of the producer;

Whereas products age in the course of time, higher safety standards are developed and the state of science and technology progresses; whereas, therefore, it would not be reasonable to make the producer liable for an unlimited period for the defectiveness of his product; whereas, therefore, liability should expire after a reasonable length of time, without prejudice to claims pending at law;

Whereas, to achieve effective protection of consumers, no contractual derogation should be permitted as regards the liability of the producer in relation to the injured person;

Whereas under the legal systems of the Member States an injured party may have a claim for damages based on grounds of contractual liability or on grounds of non-contractual liability other than that provided for in this Directive; insofar as these provisions also serve to attain the objective of effective protection of consumers, they should remain unaffected by this Directive; whereas, insofar as effective protection of consumers in the sector of pharmaceutical products is already also attained in a Member State under a special liability system, claims based on this system should similarly remain possible;

Whereas, to the extent that liability for nuclear injury or damage is already covered in all Member States by adequate special rules, it has been possible to exclude damage of this type from the scope of this Directive;

Whereas, since the exclusion of primary agricultural products and game from the scope of this Directive may be felt, in certain Member States, in view of what is expected for the protection of consumers, to restrict unduly such protection, it should be possible for a Member State to extend liability to such products;

Whereas, for similar reasons, the possibility offered to a producer to free himself from liability if he proves that the state of scientific and technical knowledge at the time when he put the product into circulation was not such as to enable the existence of a defect to be discovered may be felt in certain Member States to restrict unduly the protection of the consumer; whereas it should therefore be possible for a Member State to maintain in its legislation or to provide by new legislation that this exonerating circumstance is not admitted; whereas, in the case of new legislation, making use of this derogation should, however, be subject to a Community stand-still procedure, in order to raise, if possible, the level of protection in a uniform manner throughout the Community;

Whereas, taking into account the legal traditions in most of the Member States, it is inappropriate to set any financial ceiling on the producer's liability without fault; whereas, insofar as there are, however, differing traditions, it seems possible to admit that a Member State may derogate from the principle of unlimited liability by providing a limit for the total liability of the producer for damage resulting from a death or personal injury and caused by identical items with the same defect, provided that this limit is established at a level sufficiently high to guarantee adequate protection of the consumer and the correct functioning of the common market;

Whereas the harmonization resulting from this Directive cannot be total at the present stage, but opens the way towards greater harmonization; whereas it is therefore necessary that the Council receive at regular intervals reports from the Commission on the application of this Directive, accompanied, as the case may be, by appropriate proposals;

Whereas it is particularly important in this respect that a re-examination be carried out of those parts of the Directive relating to the derogations open to the Member States, at the expiry of a period of sufficient length to gather practical experience on the effects of these derogations on the protection of consumers and on the functioning of the common market,

HAS ADOPTED THIS DIRECTIVE:

Article 1
The producer shall be liable for damage caused by a defect in his product.

Article 2
For the purpose of this Directive 'product' means all movables, with the exception of primary agricultural products and game, even though incorporated into another movable or into an immovable. 'Primary agricultural products' means the products of the soil, of stock-farming and of fisheries, excluding products which have undergone initial processing. 'Product' includes electricity.

Article 3
1. 'Producer' means the manufacturer of a finished product, the producer of any raw material or the manufacturer of a component part and any person who, by putting his name, trademark, or other distinguishing feature on the product presents himself as its producer.
2. Without prejudice to the liability of the producer, any person who imports into the Community a product for sale, hire, leasing or any form of distribution in the course of his business shall be deemed to be a producer within the meaning of this Directive and shall be responsible as a producer.
3. Where the producer of the product cannot be identified, each supplier of the product shall be treated as its producer unless he informs the injured person, within a reasonable time, of the identity of the producer or of the person who supplied him with the product. The same shall apply, in the case of an imported product, if this product does not indicate the identity of the importer referred to in paragraph 2, even if the name of the producer is indicated.

Appendix 1: EEC Directive on strict product liability

Article 4

The injured person shall be required to prove the damage, the defect and the causal relationship between defect and damage.

Article 5

Where, as a result of the provisions of this Directive, two or more persons are liable for the same damage, they shall be liable jointly and severally, without prejudice to the provisions of national law concerning the rights of contribution or recourse.

Article 6

1. A product is defective when it does not provide the safety which a person is entitled to expect, taking all circumstances into account, including:
(a) the presentation of the product;
(b) the use to which it could reasonably be expected that the product would be put;
(c) the time when the product was put into circulation.
2. A product shall not be considered defective for the sole reason that a better product is subsequently put into circulation.

Article 7

The producer shall not be liable as a result of this Directive if he proves:
(a) that he did not put the product into circulation; or
(b) that, having regard to the circumstances, it is probable that the defect which caused the damage did not exist at the time when the product was put into circulation by him or that this defect came into being afterwards; or
(c) that the product was neither manufactured by him for sale or any other form of distribution for economic purpose nor manufactured or distributed by him in the course of his business; or
(d) that the defect is due to compliance of the product with mandatory regulations issued by the public authorities; or
(e) that the state of scientific and technical knowledge at the time when he put the product into circulation was not such as to enable the existence of the defect to be discovered; or
(f) in the case of a manufacturer of a component, that the defect is attributable to the design of the product in which the component has been fitted or to the instructions given by the manufacturer of the product.

Article 8

1. Without prejudice to the provisions of national law concerning the right of contribution or recourse, the liability of the producer shall not be reduced when the damage is caused both by a defect in the product and by the act or omission of a third party.
2. The liability of the producer may be reduced or disallowed when, having regard to all the circumstances, the damage is caused both by a defect in the product and by the fault of the injured person or any person for whom the injured person is responsible.

Article 9

For the purpose of Article 1, 'damage' means:
(a) damage caused by death or by personal injuries;
(b) damage to, or destruction of, any item of property other than the defective product itself, with a lower threshold of 500 ECU, provided that the item of property

(i) is of a type ordinarily intended for private use or consumption, and
(ii) was used by the injured person mainly for his own private use or consumption.
This Article shall be without prejudice to national provisions relating to nonmaterial damage.

Article 10

1. Member States shall provide in their legislation that a limitation period of three years shall apply to proceedings for the recovery of damages as provided for in this Directive. The limitation period shall begin to run from the day on which the plaintiff became aware, or should reasonably have become aware, of the damage, the defect and the identity of the producer.
2. The laws of Member States regulating suspension or interruption of the limitation period shall not be affected by this Directive.

Article 11

Member States shall provide in their legislation that the rights conferred upon the injured person pursuant to this Directive shall be extinguished upon the expiry of a period of ten years from the date on which the producer put into circulation the actual product which caused the damage, unless the injured person has in the meantime instituted proceedings against the producer.

Article 12

The liability of the producer arising from this Directive may not, in relation to the injured person, be limited or excluded by a provision limiting his liability or exempting him from liability.

Article 13

This Directive shall not affect any rights which an injured person may have according to the rules of the law of contractual or non-contractual liability or a special liability system existing at the moment when this Directive is notified.

Article 14

This Directive shall not apply to injury or damage arising from nuclear accidents and covered by international conventions ratified by the Member States.

Article 15

1. Each Member State may:
(a) by way of derogation from Article 2, provide in its legislation that within the meaning of Article 1 of this directive 'product' also means primary agricultural products and game;
(b) by way of derogation from Article 7(e), maintain or, subject to the procedure set out in paragraph 2 of this Article, provide in its legislation that the producer shall be liable even if he proves that the state of scientific and technical knowledge at the time when he put the product into circulation was not such as to enable the existence of a defect to be discovered.
2. A Member State wishing to introduce the measure specified in paragraph 1(b) shall communicate the text of the proposed measure to the Commission. The Commission shall inform the other Member States thereof.

The Member State concerned shall hold the proposed measure in abeyance for nine months after the Commission is informed and provided that in the meantime the Commission has not submitted to the Council a proposal amending this Directive on the relevant matter. However, if within three months of receiving the said information, the Commission does not advise the Member State concerned that it intends submitting such a proposal to the Council, the Member State may take the proposed measure immediately.

If the Commission does submit to the Council such a proposal amending this Directive within the aforementioned nine months, the Member State concerned shall hold the proposed measure in abeyance for a further period of eighteen months from the date on which the proposal is submitted.

3. Ten years after the date of notification of this Directive, the Commission shall submit to the Council a report on the effect that rulings by the courts as to the application of Article 7(e) and of paragraph 1(b) of this Article have on consumer protection and the functioning of the common market. In the light of this report, the Council acting on a proposal from the Commission and pursuant to the terms of Article 100 of the Treaty, shall decide whether to repeal Article 7(e).

Article 16

1. Any Member State may provide that a producer's total liability for damage resulting from a death or personal injury and caused by identical items with the same defect shall be limited to an amount which may not be less than 70 M ECU.

2. Ten years after the date of notification of this Directive, the Commission shall submit to the Council a report on the effect on consumer protection and the functioning of the common market of the implementation of the financial limit on liability by those Member States which have used the option provided for in paragraph 1. In the light of this report, the Council, acting on a proposal from the Commission and pursuant to the terms of Article 100 of the Treaty, shall decide whether to repeal paragraph 1.

Article 17

This Directive shall not apply to products put into circulation before the date on which the provisions referred to in Article 19 enter into force.

Article 18

1. For the purposes of this Directive, the ECU shall be that defined by Regulation (EEC) No 3180/78,[4] as amended by Regulation (EEC) No 2626/84.[5] The equivalent in national currency shall initially be calculated at the rate obtaining on the date of adoption of this Directive.

2. Every five years the Council, acting on a proposal from the Commission, shall examine and, if need be, revise the amounts in this Directive, in the light of economic and monetary trends in the Community.

Article 19

1. Member States shall bring into force, not later than three years from the date of notification of this Directive, the laws, regulations and administrative provisions necessary

---

[4] OJ No L 379, 30.12.1978, p. 1.
[5] OJ No L 247, 16.9.1984, p. 1.

to comply with this Directive. They shall forthwith inform the Commission thereof.[6]
2. The procedure set out in Article 15(2) shall apply from the date of notification of this Directive.

Article 20

Member States shall communicate to the Commission the texts of the main provisions of national law which they subsequently adopt in the field governed by this Directive.

Article 21

Every five years the Commission shall present a report to the Council on the application of this Directive and, if necessary, shall submit appropriate proposals to it.

Article 22

This Directive is addressed to the Member States.
Done at Brussels,
For the Council
The President

---

[6] This Directive was notified to the Member States on 30 July 1985.

APPENDIX 2
Sources of further assistance

Organisations

ACOT
Frascati Road
Blackrock
Dublin 18
Tel: (01) 885361

Agricultural and Food Science Research Centre
Newforge Lane
Belfast BT9 5NW
Tel: (084) 661166

American Society for Quality Control
230 West Wells Street
Milwaukee
Wisconsin 53203
USA
Tel: (16-1-414) 272-8575

An Bord Bainne
Grattan House
Lower Mount Street
Dublin 2
Tel: (01) 785788

An Bord Iascaigh Mhara
Crofton Road
Dun Laoghaire
Co. Dublin
Tel: (01) 804411

An Foras Taluntais
19 Sandymount Avenue
Ballsbridge
Dublin 4
Tel: (01) 688188

AnCO — The Industrial Training Authority
Baggot Buildings
Baggot Street
Dublin 4
Tel: (01) 685777

Bord Failte
Baggot Street Bridge
Dublin 2
Tel: (01) 765871

British Standards Institution
PO Box 104
Maylands Avenue
Hemel Hempstead
Herts HP2 4ST
England
Tel: (030442) 47688

CBF — Irish Livestock and Meat Board
Clanwilliam Court
Lower Mount Street
Dublin 2
Tel: (01) 685155

CERT
1 Ailesbury Road
Ballsbridge
Dublin 4
Tel: (01) 693522

Confederation of Irish Industries
Confederation House
Kildare Street
Dublin 2
Tel: (01) 779801

Córas Tráchtála
Merrion Hall
Strand Road
Sandymount
Dublin 4
Tel: (01) 695011

Dairy and Food Science Faculty
University College
Cork
Tel: (021) 276871

Department of Agriculture
Central Meat Control Laboratory
Abbotstown
Castleknock
Co. Dublin
Tel: (01) 213041

Department of Agriculture
Dundonald House
Belfast BT4 3SB
Tel: (084) 650111

Department of Economic Development
Netherleigh
Massey Avenue
Belfast BT4 2JP
Tel: (084) 63244

Department of Environment
Custom House
Dublin 1
Tel: (01) 742961

Department of the Environment
Stormont
Belfast BT4 3SS
Tel: (084) 63210

Department of Health
Custom House
Dublin 1
Tel: (01) 742961

Department of Industry, Trade,
Commerce and Tourism
Kildare Street
Dublin 2
Tel: (01) 614444

Department of Mechanical and
Industrial Engineering
Queen's University of Belfast
Ashby Buildings
Stranmillis Road
Belfast BT9 5AH
Tel: (084) 65171

Director of Environmental Health
Services
Belfast City Council
PO Box 234
City Hall
Belfast BT1 5GS
Tel: (084) 220202

Engineering Industries Training Board
Swinson House
Glenmount Road
Church Road
Newtownabbey BT36 7LH
Tel: (0848) 65171

Environmental Health Officers
Association
9 Aston Quay
Dublin 2
Tel: (01) 772258

European Organisation for Quality
Control
PO Box 2613
CH-3001 Berne
Switzerland
Tel: (16-41-31) 220382

Industrial Development Authority
Wilton Park House
Wilton Place
Dublin 2
Tel: (01) 686633

Industrial Development Board for
Northern Ireland
64 Chichester Street
Belfast BT1 4JX
Tel: (084) 233233

Industrial Science Division
17 Antrim Road
Lisburn BT28 3AL
Tel: (08462) 5161

Institute of Food Science and
Technology of Ireland
Sandymount Avenue
Dublin 4
Tel: (01) 688188

Appendix 2: Sources of further assistance

Institute for Industrial Research and Standards
Ballymun Road
Dublin 9
Tel: (01) 370101

Institute of Quality Assurance
54 Princes Gate
Exhibition Road
London SW7 2PG
England
Tel: 031-584-9026

Irish Goods Council
Ireland House Trade Centre
Strand Road
Dublin 4
Tel: (01) 696011

Irish Hotel and Catering Institute
11 Herbert Street
Dublin 2
Tel: (01) 763307

Irish Productivity Centre
IPC House
Shelbourne Road
Ballsbridge
Dublin 4
Tel: (01) 686244

Irish Quality Control Association
Shelbourne House
Shelbourne Road
Ballsbridge
Dublin 4
Tel: (01) 683311

Irish Society of Food Hygiene Technology
PO Box 121
Belfast BT5 5LH
Tel: (084) 661166

Irish Veterinary Association
53 Lansdowne Road
Ballsbridge
Dublin 4
Tel: (01) 685263

Kilkenny Design Workshops Ltd
Castle Yard
Kilkenny
Tel: (056) 22118

Loughry College of Agriculture and Food Technology
Cookstown
Co. Tyrone BT80 9AA
Tel: (0806487) 62491

National Board for Science and Technology
Shelbourne House
Shelbourne Road
Ballsbridge
Dublin 4
Tel: (01) 683311

National Standards Authority of Ireland
Ballymun Road
Dublin 9
Tel: (01) 370101

Office of Consumer Affairs
13 Hume Street
Dublin 2
Tel: (01) 613399

Quality Assurance Research Centre
Department of Industrial Engineering
University College
Galway
Tel: (091) 24411

School of Engineering
Regional Technical College
Sligo
Tel: (071) 3261

Shannon Development
Town Centre
Shannon
Co. Clare
Tel: (061) 61555

Technology Unit
University of Ulster
Carrickfergus Industrial Centre
75 Belfast Road
Carrickfergus BT38 8BX
Tel: (08493) 64686

Údarás na Gaeltachta
Na Forbacha
Gaillimh
Tel: (091) 21011

Union of Japanese Scientists and Engineers
5-10-11 Sendagaya, Shibuya–Ku
Tokyo 151
Japan
Tel: 16-81-3-352-2231

Directories

The following directories published by

The National Board for Science and Technology
Shelbourne House
Shelbourne Road
Ballsbridge
Dublin 4
Tel: (01) 683311

should provide additional information.

● *Science and Engineering Expertise in Irish Higher Education*–a directory of the research and professional interests of personnel in the science and technology departments of Irish colleges

● *Expertise Ireland*–a directory of Irish expertise for technical cooperation and development

● *State Analytical Services*–a directory of 102 laboratories and analytical services

APPENDIX 3
A typical quality audit check-list for a manufacturing firm

1. QUALITY PLANNING
1.1 SPECIFICATIONS/DOCUMENTATION
1.1.1 Specifications Exist for
— Raw Materials
— Intermediates
— Finished Products
(Company, Legal, Safety Requirements)
1.1.2 Documented Sampling/Inspection/Test Procedures
— Raw Materials
— Intermediates
— Finished Products
1.1.3 Documented Manufacturing Instructions
1.1.4 Approval System for Above Documentation/Specifications Adequate
1.1.5 System Adequate to Control Changes to Documentation/Specifications

1.2 EQUIPMENT (PRODUCTION AND INSPECTION)
1.2.1 Measurement Equipment Satisfactory for Intended Use and Measurement
1.2.2 Existence of Calibration Programme (Including all Necessary Equipment and Record System)
— Laboratory
— Plant
1.2.3 Method of Calibration Adequate
1.2.4 Calibration Tolerances Defined
1.2.5 Frequency of Calibration Defined and in Line with Equipment Needs
1.2.6 Equipment Tagged with Status
1.2.7 Equipment Present for Performing Necessary Calibration
1.2.8 Adequate Standards Traceable to Official Standard
1.2.9 Equipment Maintenance Schedule

1.3 QUALITY AUDIT
1.3.1 Quality Audit Carried Out
1.3.2 Scope of Audit Sufficiently Covered
1.3.3 Follow-Up After Audit Adequate

1.4 PRODUCT/PACKAGING DESIGN
1.4.1 Quality Control Analysis of Product Design
1.4.2 System for Introduction of New/Modified Products and/or Processes
1.4.3 Marketing Analysis of Product Design
1.4.4 Quality of Presentation

1.5 THIRD PARTY MANUFACTURE CONTROL (CONTRACT MANUFACTURING)
1.5.1 Operation Audited
1.5.2 Documented Procedure for Manufacturing Control

2. INCOMING MATERIAL
2.1 Documented Quality Requirements Provided to Vendor

2.2 Appraisal of Vendor Quality Capability Facility
2.3 Identification of Raw Materials
2.4 Sampling Satisfactory and as Per Procedure
2.5 Incoming Raw Materials Adequately Inspected/Tested Records Kept
2.6 System to Prevent Material Usage Prior to Inspection
2.7 Handling Adequate
2.8 Storage Conditions Adequate — *Stock Rotation System*
2.9 Outside Warehousing

3. MANUFACTURING CONTROL
3.1 Critical Control Points Established
3.2 Material Adequately Identified Quality Control Status
3.3 Sampling Satisfactory and as Per Procedure
3.4 Material Adequately Inspected/Tested
3.5 Non-conforming Material System Present and Operating
3.6 Segregation of Accept/Reject Product
3.7 Material Flow
3.8 Handling Adequate
3.9 Storage Conditions Adequate — *Stock Rotation System*
3.10 Traceability Present, if Applicable — Lot Numbering System
3.11 Stability Programme Reliability Testing etc.
3.12 Preservation of Product Quality
     — Packaging

4. RECORDS
4.1 Records Kept on all Inspections
     — Intermediate
     — Finished Products
4.2 Trend Data Present
4.3 Key Management Receive and Act on Data Reports

5. HYGIENE (SOME ITEMS APPLICABLE ONLY TO FOOD)
5.1 Microbial Monitoring of Air, Surfaces, Water, Adequate
5.2 Chemical Monitoring of Water Adequate
5.3 Housekeeping Programme:
     — Areas Cleaned
     — Method of Cleaning
     — Frequency of Cleaning
     — Cleaning Materials Used
5.4 Rodent Control
     Insect Control
     Dust Control
     Bacteriological Control
5.5 Equipment Cleaning Programme Cleanliness of Equipment
5.6 Facility in Good Repair (Floor, Walls, Door)
5.7 No Miscellaneous/Redundant Items Present
5.8 Adequate Handwash, Toilet Facilities
5.9 Cleanliness of Facility

Appendix 3: A typical quality audit checklist

5.10 Cleaning Materials:
- Present
- Suitable for Job, Properly Stored
- Clean

5.11 Utensils Clean and Adequately Stored to Maintain Cleanliness

5.12 Personnel Correctly Dressed for Work Being Performed, Uniforms Clean and in Repair

6. TRAINING
6.1 Training Needs Defined for Relevant Personnel
6.2 Documented Training Programmes
6.3 Training Instruction, e.g. Training Manual
6.4 Training Methods Established and Carried Out: Training Aids, e.g. Films, etc.
6.5 Training Centre
6.6 Training Records Current
6.7 On-Going Training for Relevant Personnel

7. CUSTOMER SERVICE
7.1 Documented Procedure on Complaint Handling
7.2 Record of Customer Complaints
7.3 Handling of Customer Complaints
- Person Responsible
- Action Taken
- Follow-Up

7.4 Periodic Report of Complaints to Management
7.5 Action Taken Internally to Rectify/Minimise Problem
7.6 What Procedure is Used for Field Servicing
7.7 Quality of Service

8. MANAGEMENT OF PRODUCT QUALITY
8.1 Quality Manual Present
8.2 Quality Related Goals/Objectives, Defined, Reviewed, Achieved
8.3 Quality Cost Procedure in Operation
8.4 Quality Cost Data Evaluated and Utilised
8.5 Quality Responsibility
8.6 Periodic Review of Quality Data with Plant Manager and Staff
8.7 Commitment

Source: IS 303: 1984 *Quality Audit*

Bibliography

American Institute of Baking, *Quality Assurance Manual for Food Processors*, Kansas, 1980
American Society for Quality Control, *Quality Costs — What and How*, Milwaukee, 1971
American Society for Quality Control, *Quality Control and Reliability Management*, Milwaukee, 1973
American Society for Quality Control, *Quality Circles — Applications, Tools and Theory*, Milwaukee, 1976
American Society for Quality Control, *Guides for Reducing Quality Costs*, Milwaukee, 1977
Assurance Manual for Food Processors, Kansas, 1980
AnCO, *Total Quality Control*, Dublin, 1985
Anson, C. J., *Quality Control as a Tool for Production*, London, 1980
Asian Productivity Organisation, *Japan Quality Control Circles*, Tokyo, 1972
Bajaria, Hans J., *Quality Assurance — Methods, Management and Motivation*, Michigan, 1981
Bartlett, J. B., *Success and Failure in Quality Circles*, Cambridge, 1983
Besterfield, Dale H., *Quality Control*, Englewood Cliffs, 1979
British Standards Institution, *Quality Assurance — BSI Handbook 22*, London, 1981
British Standards Institution, *Engineering Metrology*, London, 1984
Burr, Irving W., *Engineering Statistics and Quality Control*, New York, 1953
Burr, Irving W., *Statistical Quality Control Methods*, New York, 1976
Burr, Irving W., *Elementary Statistical Quality Control*, New York, 1979
Calabro, S. R., *Reliability Principles and Practices*, New York, 1962
Caplan, Frank, *The Quality System*, Pennsylvania, 1980
Caplen, R. H., *Quality and the Supervisor*, London, 1974
Caplen, R. H., *A Practical Approach to Quality Control*, London, 1978
Carroll, C., *Building Ireland's Business — Perspectives from Pims*, Dublin, 1985
Charbonneau, Harvey C. & Webster, G. L., *Industrial Quality Control*, Englewood Cliffs, 1978
Cochran, William G., *Sampling Techniques*, New York, 1977
Cooper, Murray S., *Quality Control in the Pharmaceutical Industry*, London, 1972
Cowden, Dudley J., *Statistical Methods in Quality Control*, Englewood Cliffs, 1957
Croome, D. J. & Sherratt, A. F. C., *Quality and Total Cost in Buildings and Services Design*, Lancaster, 1977
Crosby, Philip B., *Quality is Free*, New York, 1979
Curran, S. & Curnow, R., *The Penguin Computing Book*, Harmondsworth, 1983
Davies, Owen L., *The Design and Analysis of Industrial Experiments*, London, 1954
Deming, W. Edwards, *Statistical Adjustment of Data*, New York, 1943
Dennis, P. O., Blanchfield, J. R. & Warr, A. G., *Food Control in Action*, London, 1980
Desmond, David J., *Quality Control Workbook*, Farnborough, 1978
Dewar, Donald L., *The Quality Circle Guide to Participation Management*, Englewood Cliffs, 1980

Bibliography

Dixon, Wilfred J. & Massey, Frank J., *Introduction to Statistical Methods*, 3rd ed., New York, 1969

Dodge, H., & Frend Romig, Harry G., *Sampling Inspection Tables, Single and Double Sampling*, New York, 1959

Drury, C. G. & Fox, J. G., *Human Reliability in Quality Control*, London, 1975

Duncan, Acheson J., *Quality Control and Industrial Statistics*, 4th ed., Homewood, Illinois, 1974

Elvy, B. Howard, *Marketing Made Simple*, London, 1972

Enrick, N. L., *Quality Control and Reliability*, New York, 1977

Eyre, E. C., *Effective Communication — Made Simple*, London, 1983

Feigenbaun, Armand V., *Total Quality Control*, 3rd ed., New York, 1983

Fisher, Sir Ronald A., *Statistical Methods for Research Workers*, 14th ed., New York, 1970

Fitzpatrick, J. & Kelly, J., *Perspectives on Irish Industry*, Dublin, 1985

Frechette, V. D., Pye, L. D. & Rase, D. E., *Quality Assurance in Ceramic Industries*, New York, 1979

Freund, John E., *Mathematical Statistics*, Englewood Cliffs, 1962

Goldmann, Heinz M., *How to Win Customers*, London, 1980

Goldsmith, W. & Clutterbuck, D., *The Winning Streak*, Middlesex, 1954

Goodenough, P. W. & Atkin, R. K., *Quality in Stored and Processed Vegetables and Fruit*, London, 1981

Grant, Eugene L. & Leavenworth, Richard S., *Statistical Quality Control*, 5th ed., New York, 1979

Green, A. E. & Bourne, A. J., *Reliability Technology*, Chichester, 1978

Guenther, William C., *Sampling Inspection in Statistical Quality Control*, New York, 1977

Hald, A., *Statistical Theory of Sampling Inspection by Attributes*, London, 1981

Hansen, Bertrand L., *Quality Control: Theory and Applications*, Englewood Cliffs, 1963

Herschdoerfer, S. M., *Quality Control in the Food Industry*, London, 1980

Hines, W. W. & Montgomery, D. C., *Probability and Statistics in Engineering & Management Science*, New York, 1980

HMSO, *Standards Quality and International Competitiveness*, London, 1982

Hoel, P. C., *Introduction to Mathematical Statistics*, 4th ed., New York, 1971

Ingle, S., *Quality Circles Master Guide*, Englewood Cliffs, 1982

Inhorn, Stanley L., *Quality Assurance Practices for Health Laboratories*, Washington, DC, 1978

Ishikawa, Kaoru, *Guide to Quality Control*, Tokyo, 1981

Jardine, A. K. S., Macfarlane, J. D. & Greensted, C. S., *Statistical Methods for Quality Control*, London, 1975

Johnson, A. H., & Peterson, M. S., *Encyclopedia of Food Technology*, Connecticut, 1974

Juran, J. M., *Managerial Breakthrough*, New York, 1964

Juran, J. M., *Quality Control Handbook*, 3rd ed., New York, 1974

Juran, J. M., *Upper Management and Quality*, New York, 1983

Juran, J. M. & Gryna, Frank M. Jr., *Quality Planning and Analysis*, New York, 1980

Kefner, Trevor, *Problem Analysis and Decision Making*, New Jersey, 1979

Kramer, A. & Twigg, B. A., *Quality Control for the Food Industry*, Connecticut, 1980

Lapedes, Daniel N., *Directory of Scientific and Technical Terms*, New York, 1978

Lee, Alec M., *Systems Analysis Frameworks*, London, 1970

Lester, R. H., Enrich, N. L. & Mottley, M. E., *Quality Control for Profit*, New York, 1977

Lewis, S. M. & Coster, J. F., *Quality Control in Haematology*, London, 1975

McCall, J. L. & French, P. M., *Metallography as a Quality Control Tool*, New York, 1980

McElearney, John, 'The Influence of Computer Systems in the Quality Control Function', dissertation, Dublin, 1985
Montgomery, D. C., *Design and Analysis of Experiments,* New York, 1976
Mood, A. M., & Graybill, F. A., *Introduction to the Theory of Statistics,* New York, 1963
Murphy, John A., *Hygiene in Practice,* Dublin, 1985
National Standards Authority of Ireland, *Quality Assurance Standards Handbook,* Dublin, 1984
Ott, Ellis R., *Process Quality Control,* New York, 1975
Paget, G. E., *Quality Control in Toxicology,* Lancaster, 1977
Parker, Maurice, *Manual of British Standards in Engineering Drawing,* London, 1984
Parry, V. G., *The Control of Quality,* London, 1973
Peters, T. J. & Waterman, R. H., *In Search of Excellence,* New York, 1982
Peterson, M. S. & Johnson, A.H., *Encyclopedia of Food Science,* Connecticut, 1978
Price, F., *Right First Time — Using Quality Control for Profit,* Hampshire, 1984
Robertson, A. G., *Quality Control and Reliability,* London, 1971
Roche, J. G., 'Strict Product Liability has Arrived', *Management,* Jan. 1986, Galway
Sasaki, N. & Hutchins, D., *The Japanese Approach to Product Quality,* Oxford, 1984
Sayle, Allan J., *Management Audits,* London, 1981
Sharpe, R. S., Cole, H. A. & West, J., *Quality Technology Handbook,* Surrey, 1978
Shaw, John C., *The Quality Productivity Connection,* New York, 1978
Steel, Robert G. & James, H. Torrie, *Principles and Procedures of Statistics,* New York, 1960
Thorner, M. E. & Manning, P. B., *Quality Control in Food Service,* Connecticut, 1976
Vaughan, Richard C., *Quality Control,* Ames, Iowa, 1974
White, L.T., *Strengthening Small Business Management,* Washington, DC, 1971
Whitehead, T. P., *Quality Control in Clinical Chemistry,* New York, 1977
Wilson, N. R. P., *Meat and Meat Products, Factors Affecting Quality Control,* London, 1981

Index

acceptable quality levels (AQLs) 31, 37, 199
acceptable sampling 196-7
advertising *see* product information
Aer Lingus 79
American Food and Drugs Administration 47
American Society for Quality Control 71
AnCO 116, 117, 118, 121, 149
ANSI N 45.2 30, 165
approval marks 161
assembly systems 40
automated production 40-1
averages 178-80, 190
    control chart 206-7

Bailey, R. & A. 2, 10, 32, 59, 88, 155
    complaints system 74-5
    labels 107
    out of specification report 202
    quality manual 136
    quality policy 12
    shipping 53-4
    tasting 158
Ballyclough Co-operative Creamery 31
Ballyfree Farms 51, 115
bar charts 175, 176
batch records 35, 201-3
batch testing 31, 34
'bathtub' curve 21, 22
Bell Telephone Co. 64, 178
bell-wether projects 15-16, 17, 141
bi-modal 179-80
bin systems 45, 60
binomial distribution 186-7
body language 107-8
Braun 32, 192
BS (British Standards) 22, 157, 161
    BS 5750 22, 27, 30, 148, 165
    BS 6143 224
buying 27-35
    influences 28
    supplier assessment 16, 27-30, 35
    supplier relations 30-4

calibration 154-5, 245
Camp's formula 60-1
Caplan, Frank 14
Carney, P. 77

cause and effect charts 169, 170-1
CEN 159
CENELEC 159
central limit theorem 189, 229
central location product test 234
Charge of the Light Brigade 69
check sheets 169-70
Chemoflon 42, 201, 203
CMP Dairy 101, 106
common law
    and the consumer 238
communication 105-13
    types of 105-9
    written 109-13
    see also information quality standards
complaints 9, 71-6, 83, 118, 246
    Aer Lingus 79
    computers and 216
    corrective action 75
    handling of 12, 72-6
    input system 74
    response system 75
    system management 75-6
computer-aided manufacture (CAM) 40
computers 64-5, 93, 108, 209-18
    input/output 211-13
    in production 217-18
    programs 213-15
    in quality assurance 215-17
    structure of 209-11
Confederation of Irish Industry (CII) 149
consumers *see* customers
Consumer Information Act 237, 239-40
Consumer Products Safety Commission 242, 246
consumer protection 200, 236-41
    in Ireland 237-40
Consumers' Association 237
Consumers' Association of Ireland 82, 237, 240
content theories of motivation 97-9
control charts 175-7, 200-8
    procedure 203-6
    types of 206-8
controlled environment area 50
Córas Tráchtála 26
corporate culture 3, 100-1

Index

Corrective Action Requests 75, 147, 148
cumulative sum control charts 208
customers 4, 135, 228-35
   in-depth research 233
   panels 231-3
   and quality 71-6
   and service industries 82-3
   surveys of 229-31
   see also market research

dairy industry 1, 91
Dairyland 40, 54
Data Products 60
defects
   control charts for 207
design 18-26
   life-cycle costs 18
   qualities of 19-25
Digital Equipment 30
documentation 17, 55, 200-8
   and liability 246-7
Drogheda and Dundalk Dairies 44-5
drugs 38, 47, 48, 200; see also Good
   Manufacturing Practice

80/20 analysis 171-4, 182, 186
economic batch quantity (EBQ) 60-1
economics see quality costs
entropy 87, 94
Eurolift 39
European Economic Community (EEC) 238,
   240-1
   Directive 248-54
   food additives 47-8
   product liability 242, 244
exponential distribution 21
extreme value distribution 21-2

failures 21-2, 36
   failure mode and effect analysis (FMEA)
   24, 245
feedback control 87, 89, 237
field monitoring 246
final check systems 38-40
Fitzwilliam Quality Assurance 145
FMC International 28-9
Fog Index 111-12
food 38, 47-8, 200
   additives 47
   hygiene 52
   see also Good Manufacturing Practice
frequency distribution 184-6

Good Manufacturing Practice (GMP) 38,
   47-56
   buildings and facilities 50-1
   components 52
   equipment 51-2
   laboratory controls 54

   organisation and personnel 50
   packaging 53
   production and process controls 52-3
   records and reports 55
   warehousing and distribution 53-4
Guinness, Arthur 1
Gunning, Robert 111

Hawthorne factory experiments 96
Herzberg, Dr Frederick 98
histograms 175, 176, 185, 186, 189
Howmedica 125, 194

in-process control 36-46
   job shop production control 37-8
   mass production systems 38-41
   operator control 41-5
industrial market research 235
Industrial Research and Standards Act 239
Information quality standards 64-70
   channel 66-8
   decoder 69
   encoder 65-6
   message 66
   noise 68
   receiver 69-70
   source 65
inspection bay 31
inspection procedures 11, 31, 34, 35, 36,
   191-9, 245
   acceptance sampling 196-9
   errors 195-6
   four points of 191-3
Institute of Industrial Research and Standards
   (IIRS) 149, 159, 161, 162
   weights 153-4, 239
instructions see product information
intermediate check systems 38
International Electrotechnical Commission
   (IEC) 159
International Organisation for
   Standardization (ISO) 159
investment 14-15
Irish Congress of Trade Unions 149
Irish Fher Laboratories 137-8
Irish Quality Control Association 116, 121,
   143-4, 162
   Quality Mark Scheme 148-9
IS (Irish Standards) 1
   IS 257 Quality Glossary 161
   IS 300 Quality System 13, 27, 30, 142,
   143, 148, 159, 161-5
   management 162-4
   requirements 164-5
   IS 301 Sampling Procedures 162, 199
   IS 302 Quality Manual Preparation
   Guide 162
   IS 303 Quality Audit 143, 162
   IS 304 Guide to Quality Costs 162, 224

Index

IS 305 Guide to Quality Management in the Service Industries 162
Ishikawa, Dr Kaoru 123, 158, 161, 174
    diagrams 16, 167, 170

job enrichment schemes 104
job shop production control 37-8
Juran, Dr J.M. 195
Just-In-Time production 29, 34, 57-63
    reducing setup cost 60-1
    statistics of production 61
    suppliers and 62

kanban system 60
Kelvin, Lord 156
Kerrygold 1
key quality characteristics 9, 11, 28, 78-82
Kilkenny Design Workshops 19, 26, 150

labelling *see* product information
language 108; *see also* communication
liability *see* product liability
lognormal distribution 22

McGregor, Douglas 97-8
machine capabilities 16
maintainability 24-5
management 3-4, 7-9, 14, 15, 17, 100-1
    competence of 103
    giving orders 101-3
    and operator control 41-3
    and quality circles 123, 124, 126
    and staff motivation 97-8, 103-4
market research 10, 73, 74, 228-35
    customer surveys 229-31
    in-depth techniques 233
    panel techniques 231-3
Marks and Spencer 71
Maslow, A.H. 96, 97, 105
mass communication 108-9
mass production 37, 38-41, 87-8, 150
Mayo, Elton 96, 100
mean 179
Mean Time 21
measurement systems 150-6
    errors 155-6
    science of 52-5
Measurex 204
median 179
Merck, Sharp and Dohme 68
metric system 151-2
metrology 150-6
    errors 155-6
    measurement units 151-2
    precision 150-1
    science of measurement 152-5
Milne, A.A. 94
Mitchelstown 49, 72
mode 179-80

motivation 14, 76, 95-104, 120
    hierarchy of needs 96-7
    lessons of theory 103-4
    maintenance and motivating factors 98-9
    orders 101-3
    process models of 99-100
    sociology of the workplace 100-1
    in training 115, 120
Murphy Brewery 217
Murphy's Law 65, 78-9

Nader, Ralph 236-7
National Standards Authority of Ireland 159, 161, 162
non-verbal communication 106-8, 133
normal distribution curve 187-9
Nypro 23, 154, 218

operators 9, 16, 41-5
    control techniques 43-5
    performance testing 24-5
    and quality circles 123, 126, 129
    see also motivation
orders, getting and giving 101-3
output measurement 91-2

packaging *see* product information
panel surveys 231-3
Pareto, Vilfredo 171-2
    Parento analysis *see* 80/20 analysis
Pascal's triangle 197
performance, measurement of 91-2
personnel *see* management; operators
Pfizer Chemicals 55-6
PIMS database 2, 222-3, 236
(An) Post 82
precision instruments 150-1
presumptive testing 48
price
    and complaints 71, 72-3
printing companies 27, 57-8
probability 183-4
problem-solving 166-77
    solutions 167-8
    tools of 167-77
    understanding 166-7
process documents 200, 201-6
process models of motivation 99-100
production information 231
    advertising 71, 235, 246
    instructions 20, 73
    labelling 53, 245-6
    packaging 53, 245-6
product liability 7, 17, 200, 242-7
    costs 247
    EEC Directive 248-54
    implications for industry 244
    insurance 244-5

in Ireland 242-4
   prevention 245-7
product recall 246
production
   computers in 217-18
production lines 87-8
proportions 182
purchasing *see* buying

quality assurance 13-15, 27, 142, 215-17
   handbook 165
quality audits 142-9
   depth of 144
   follow-up 148
   information gathering 146-7
   Quality Mark Scheme 148-7
   report 147-8
   types of 143-4
quality circles 122-9
   components of 124-6
   introduction of 126-7
   problems of 128-9
   working of 127
quality controls 2-4, 7-8
   adjustments 93-4
   computer program for 214-15
   elements of 11, 12, 89-91
   in-process 36-46
   in service industries 78
   objectives 11-13
   sensing 91-2
   standards 92-3, 158, 161-5
   subjects 78, 89-91
quality costs 7-8, 17, 57, 221-7
   control costs 225-7
   and income 222-3
   and liability 247
quality manual 31, 133-41, 201, 208
   planning 134-8
   structure and content 138-40
Quality Mark Scheme 148-9
quality organisation 14-15
quality programme 9-17, 129
quality project team 15
Quality System Management *see* IS 300
quality systems 87-94, 200-1
   control function 89-94
   elements of 88-9
   quota samples 229-30

random samples 196-9, 230
range 180, 181
   control charts for 207
Relative Product Quality (RPQ) 2, 7, 9-11, 17, 71-6
   and income 222-3
reliability 21
retail auditing 232-3
Right First Time 41, 45

Rowntree-Macintosh 31

safety analysis 20-1, 23-5
   and product liability 242-7
Sale of Goods and Supply of Services Act 238-9
sampling 37, 48, 191-9
   acceptance sampling 196-9
   and customer surveys 229
sampling frame 230
scatter, measures of 180-2
scatter diagrams 173, 174-5
Semperit 216
service industries 77-84
   measures of quality 80-4
   quality characteristics 78-82
   sales staff 83-4
Shaw, G.B. 109
Shewhart 178
ship-to-stock system 34, 62
signs 106-7, 113
Sligo Dairies 1
SNAFU 87, 94
sociology
   of the workplace 100-1
staff *see* management; operators
standard deviation 181-2
   control chart for 207
standard operating procedures (SOPs) 56, 133
standardisation 157-65
   definition of 159-61
   importance of 157-9
   see also BS (British Standards); IS (Irish Standards)
Standex Ireland 223
statistical quality control 36, 61, 178-90
   averages 178-80
   binomial distribution 186-7
   frequency distribution 184-6
   normal distribution 187-9
   proportion 182-4
   scatter 180-2
stocks 57-61
   and Just-In-Time production 58-9, 61
stratification 171
summary records 200, 208
suppliers 1
   assessment of 16, 27-30, 35, 191, 200, 208
   and Just-In-Time production 62
   relations with 30-4
symbols 106-7
System Industries 217
systems approach 38

TARP report 73
Taylor, F.W. 95-6
Thalidomide 46

268

Index

Theory X and Theory Y 97-8
Thermoking and Digital 67
time-and-motion study 95-6
traceability 201-3
trade unions 42, 126
training programmes 14-17, 115-21
    methods 116-19
    objectives 119-21
    and quality circles 123-4
    in service industries 84

University College, Galway 3
US Department of Defense 22
usability 20

vendors *see* suppliers
Virginia Milk Products 32

Wang 211
Wavin Pipes 193
Weibull distribution 22
Westinghouse factor 172
Whitney, Eli 150
work-in-progress 58
worker participation 104
workplace, sociology of 100-1
written language 109-13

Yoplait 41, 232-3

Zero Defect schemes 14, 45